NEW MEXICAN TINWORK, 1840–1940

FRONTISPIECE
Photograph of Maria Lente and interior of Governor Lente's residence, Isleta Pueblo, circa 1910. Unidentified photographer. Photo courtesy Museum of New Mexico, negative number 13344.

N·E·W M·E·X·I·C·A·N
TINWORK
1840–1940

Lane Coulter
Maurice Dixon, Jr.
Foreword by Ward Alan Minge

UNIVERSITY OF NEW MEXICO PRESS
Albuquerque

This book has been supported in part by a generous grant
from the L. J. Skaggs and Mary C. Skaggs Foundation.

Library of Congress Cataloging-in-Publication Data

Coulter, Lane, 1944–
New Mexican tinwork, 1840–1940 / Lane Coulter, Maurice
Dixon, Jr. ; foreword by Ward Alan Minge.—1st ed.
p.   cm.
Includes bibliographical references.
ISBN 0–8263–1180–6
1. Tinsmithing—New Mexico—History.   2. Tinsmiths—New
Mexico—History.   3. Tinware, American—New Mexico.
4. New Mexico—Industries.   I. Dixon, Maurice, 1947–
II. Title.   III. Title: Tinwork.
TS600.C68   1990
739.5′32′09789—dc20   89–28253
CIP

CONTENTS

FOREWORD

*W*ith this beautiful publication, New Mexican tinwork becomes something far more than a subject for idle curiosity. The authors have tackled a nearly forgotten craft with incredible success. In an area where little was known or even remembered, Lane Coulter and Maurice Dixon have documented a tradition, defined styles, and introduced dating methods to describe this fragile and once popular art form, adding another volume to the available serious examinations of southwestern material culture.

The authors' interest in southwestern metalwork goes back a number of years. Both are trained studio artists with advanced degrees in art. Coulter not only trained as a metalsmith but worked at silversmithing and taught metalwork throughout the area. Dixon discovered the prevalence of tinwork in the regional folk arts of New Mexico through long association with art galleries where he handled local tinwork, notably disposing of personal pieces left by the late Robert Woodman, a Revival period tinsmith in Santa Fe. What appears providential is their meeting of minds over a building project which led to the discovery of a mutual growing interest in tinsmithing. This was further stimulated by their awareness that tin products in New Mexico indeed formed an important artistic statement and that little was really known about the genre and its creators. Fortunately, their research into collections and archives produced results for tinwork that stand with E. Boyd's similar detective work in the area of New Mexican religious artists.

The authors were plagued by many problems, but principally by the anonymity of the craftsmen. Gifted tradespeople accompanied the first colonists under Don Juan de Oñate in 1598, importing European traditions to the New World nearly 400 years ago. Metalworkers and carpenters were a part of this settlement, and many more followed to help create the necessities of colonial life. Unfortunately, preferences for European styles persisted in New Mexico and, along with these tastes, a supercilious attitude toward *efectos del pais* (a term used in the old Spanish documents for articles produced by the craftsmen). Any significance of local produce went disregarded by the literates of the early settlements, and efectos were often omitted from official documents, estate inventories, wills, and testaments. To the dismay and utter frustration of cultural historians, this preference for imports threw a blanket of neglect and ignorance over efectos; in most cases, consequently, the survivals are not linked to a given craftsperson. Rarely, if ever, did they sign or date their products. Magnificent pieces deserving of unquestionable provenance have survived undated and unsigned, thus adding a dating problem to that of anonymity. Finally, since colonists were more inclined to import objects from Mexico, Europe, and elsewhere when the opportunity arose, efectos must be carefully sorted to arrive at some idea of the quantity and quality of those produced locally.

Were the poor, then, lucky inheritors of these remarkable examples of native art? It appears certain that efectos began replacing imports generally toward the end of the eighteenth century. At least they began to appear more openly in estate papers of the period. But by the 1840s, Yankee traders found that the larger homes in Santa Fe and along the Rio Grande were being furnished with goods made in the eastern United States and either purchased through Santa Fe Trail trade or, more likely, obtained during expeditions organized by local families for trading in St. Louis and beyond. Nevertheless, the more private rooms and dependencies (outbuildings) continued to be furnished with efectos, which were an intimate and necessary part of life. Unquestionably, the poor relied almost entirely on efectos, producing much for their own use, as they were unable to afford imports even for their local village churches.

To their credit, Coulter and Dixon partially resolved these problems for New Mexican tinwork by applying several imaginative approaches to classification. They distinguished New Mexican workmanship from eastern products as well as from Mexican ones. Specific clues to dating pieces and styles became an exciting pursuit. Their descriptions of tin workshops is bound to fascinate the reader, and their definition of a classic period of tin production in New Mexico is an enormous contribution toward a better understanding of Hispanic material culture in the Southwest.

Particularly interesting, in view of recent interpretations of New Mexican efectos, is the growing evidence to support long-standing traditions in these crafts. The authors date tin production in New Mexico starting in 1840; however, such vigorous, sophisticated forms, designs, and techniques were not born there overnight. Hispanic metalworkers abounded during the Colonial period, and the advent of cheap and abundant tin, mainly brought over the Santa Fe Trail, stimulated the craftspeople in more fanciful ways than had been possible since the days of the first settlers. Their conditions for survival in this isolated region of the Spanish world dictated most uses for scarce iron which had to be imported from Mexico or Europe. Yet, the many illustrations included here reveal an astonishing range of creativity and spiritual strength inherited by their descendants.

Ward Alan Minge
Corrales, New Mexico

ACKNOWLEDGMENTS

*N*umerous people offered advice and help to us in many ways during the course of our study. We are especially indebted to Dr. Ward Alan Minge for writing the foreword, making suggestions for changes in the manuscript, and providing enthusiastic support throughout the project. Gloria Giffords also read the manuscript and made numerous suggestions for improvement for which we are grateful.

During the period of our survey, staff members from numerous museums were particularly helpful and interested, making our research more profitable. At the Museum of International Folk Art in Santa Fe we are especially thankful for the support of Charlene Cerny, Donna Pierce, Helen Lucero, Judy Sellars, Robin Farwell, Andrea Gillespie, and Joan Tafoya. We would like to thank Arthur Olivas and Richard Rudisill of the Museum of New Mexico Photo Archives for bringing the Mathews photographs of tinwork to our attention. We are also grateful to the following people in Santa Fe:

Richard Salazar and Al Regensberg of the State Archives; Orlando Romero of the Museum of New Mexico History Library; Charles Bennett at the Palace of the Governors; and Marina Ochoa of the Archdiocese of Santa Fe.

Staff members of other museums in New Mexico who helped us in our study were: Byron Johnson of the Albuquerque Museum; Jim Featherstone of the New Mexico State University Art Gallery in Las Cruces, along with Kate Wagle of the Art Department; Mary Veitch Alexander at the Gadsden Museum in Mesilla; Jack Boyer of the Kit Carson Foundation in Taos; and Ann Hedlund at the Millicent Rogers Museum and David Witt at the Harwood Foundation, both in Taos.

Katie Davis of the Colorado Historical Society in Denver was instrumental in arranging our introduction to the curators of their outlying historical sites: Josephine Lobato at Fort Garland, Carrie Kramer in Pueblo, and Joy Poole at the Baca House in Trinidad. We are grateful to Kathy Wright and Jonathan Batkin of the Taylor Museum in Colorado Springs; Paul Cordova of the A.R. Mitchell Museum and Gallery in Trinidad; and Robert Stroessner of the Denver Art Museum, all in Colorado.

In Arizona, we would especially like to thank Fred McAninch of the Tucson Museum of Art; James Griffith of the Southwest Folklore Center in Tucson; and Ann Marshall of the Heard Museum in Phoenix. In California, Claudine Scovill of the Southwest Museum in Los Angeles made available the tinwork collection housed at Casa de Adobe. Also helpful were Lonn Taylor of the Smithsonian Institution and Lauri Weitzenkorn at the Index of American Design of the National Gallery of Art, both in Washington, D.C.

A number of private individuals provided information concerning tinsmiths and tinsmithing in New Mexico. Among them are tinsmiths Emilio and Senaida Romero, Bonafacio Sandoval, and the late Pedro Quintana. Marie Romero Cash selflessly provided important information about the location of some of the workshops from her own survey of New Mexican religious art. Other people who contributed their help were the late Carmen Espinosa, Polo Gomez, Ray Herrera, Ted Jojola, Yvonne Lange, Maggie McDonald, Ben Martinez, Father Jerome Martinez y Alire, Christine Mather, Mr. and Mrs. J. Paul Taylor, and Will Wroth.

Some of the antique dealers who shared their experience and knowledge with us were Robert Ashton, Jack Calderella, Robin Cleaver, Ray Dewey, Murdoch Finlayson, Robert Gallegos, Tony Garcia, Victor Hansen, John Hill, Will Knappen, Cam Martin, Rowena Martinez, Reggie Sawyer, and Pauline Stevens.

Without the many private collectors who opened their homes and collections to us, our survey would be incomplete. These collectors are acknowledged in the photo captions throughout the book.

Four people to whom we are especially grateful are Martha Lindsey for many patient hours spent at the word processor, Sarah Nestor for her work in editing the manuscript, and Roderick Hook and Fred Sisneros for their work on the photographs. Additionally, we thank our editor at the University of New Mexico Press, Dana Asbury, and Beth Hadas, director of the press, for their patience. We are indebted to the Museum of International Folk Art Foundation for their support of the photographic work for the book. We must also thank our friends and families for their help and patience during the last three years. And finally, we must acknowledge all of the New Mexican tinsmiths, known and unknown, whose work was the impetus and basis for our study.

▼▼▼▼▼▼▼▼▼▼▼▼▼▼▼▼▼▼▼▼▼▼▼▼▼▼▼▼▼▼▼▼
INTRODUCTION

*W*hen we initiated this study in 1985, we intended to research nineteenth-century tinwork from all the southwestern states settled by the Spanish—Texas, Colorado, Arizona, and California as well as New Mexico. Although metalwork (primarily iron and copper) was produced by Hispanic craftsmen throughout the Southwest, we have discovered that decorative tinwork was strictly a New Mexican phenomenon. This resulted from a simultaneous exposure of the metalsmiths to the materials and culture of both the United States and Mexico in the mid-nineteenth century. Tinwork prototypes very likely came up the Chihuahua Trail from Mexico, while tin cans, wallpaper, and glass became available over the Santa Fe Trail from the East. Our studies have shown that all of the tinwork in Colorado collections was made in New Mexico and that little or no documented tin has been collected in Texas or California. Mid-nineteenth-century paintings of interiors of Spanish homes in San Antonio show numerous wall sconces, although it is not possible to determine whether they were made of tin or wood.[1] One brief reference to tinwork made in Arizona mentions a tin candle sconce in Tucson in 1870.

Although often overlooked by collectors and scholars, Hispanic tinwork was an important part of the religious folk crafts of New Mexico during the last half of the nineteenth century. While wood-working, weaving, and blacksmithing were all developed by the end of the eighteenth century, tinworking was the last Spanish craft form to evolve. This late development was a result of the extremely limited availability of materials until after the American occupation of New Mexico by the U.S. Army in 1846. While tinwork flourished in the last half of the nineteenth century, the quantity of work by native *santeros* (carvers of saints and religious images) gradually declined. Wooden *retablos* (religious paintings) were replaced by tin-framed French and German engravings and lithographs that were often dispensed by the French-born priests brought to New Mexico by Bishop Lamy after 1851. Thus, New Mexican tinwork functioned as a bridge between the religious art of the Spanish Colonial period and the coming modernization of the twentieth century.

The initial flourish of New Mexican tinsmithing spans a rather brief period of only seventy-five years, from about 1840 to 1915. The nineteenth-century craft was supplanted after World War I by the Spanish Pueblo Revival movement, which was promoted by Anglo artists intrigued with the Spanish and Indian architecture and crafts they found in the Southwest.[2] Objects traditionally made by

Hispanic tinsmiths were no longer needed by a society served by daily railroad service from all parts of the country. Electric lights replaced the need for candle sconces in all but the most remote mountain areas. Ready-framed chromolithographs of religious subjects were commonly available, and plaster statues of saints took the place of old hand-carved *bultos* (three-dimensional religious carvings) made by native santeros. The Revival tinsmiths, in response, developed new forms of tinwork to serve the needs of the twentieth-century market.

This book examines a broad cross section of New Mexican tinwork within the context of the Catholic culture of late-nineteenth-century Spanish New Mexico as well as Revival period and Mexican tinwork. We will show that tinwork was the product of skilled professionals, not the work of amateurs. The tinsmiths were familiar with the work of their peers and quickly developed designs that are unique in American folk art.

Very little has been written about the history of tinwork in New Mexico; however, the major efforts are worth noting. Mary Austin wrote briefly about tinwork in her article, "Spanish Colonial Furnishings in New Mexico," in *Antiques* magazine in 1933. In her foreword to *Tin Craft in New Mexico*, a Revival period workbook published by the State Department of Vocational Education in 1937, Carmen Espinosa provided much of the initial research about the history and use of tinwork in New Mexico. *The Index of American Design*, a WPA project of the Depression era, produced twenty-eight watercolor renderings of tinwork in its extensive survey of New Mexican folk arts between 1936 and 1938. In addition, artists of *The Southern California Project* of the *Index* made seven renderings of tinwork from the Felipe Delgado Estate, now in the Southwest Museum in Los Angeles. In his pioneering *New Mexico Village Arts* published in 1949, Roland Dickey devoted seven pages to a generalized study of the popular use of native tinwork.

The most extensive work has been done by the late scholar E. Boyd, who wrote several articles in the 1940s and 1950s and included four pages with illustrations devoted to tinwork in her classic book, *The Popular Arts of New Mexico*.[3] A few scattered articles from the Revival period of the 1930s and some more recent pictorial articles make up the remainder of works written about New Mexican tinwork.

For our study, we have surveyed virtually all major public and private collections in the Southwest and have assembled photographs of over 1,000 pieces of tinwork. An early impetus to the research was the discovery of 150 photographs of tinwork by the late Santa Fe architect Truman Mathews.[4] These photographs were probably taken in the mid-1930s during his research for the measured drawings he produced for the state vocational handbook; however, the photographs unfortunately had no accompanying information concerning the size, provenance, or whereabouts of the pieces. Some of the tinwork represented in Mathews's photographs is now in the collections of the Museum of International Folk Art in Santa Fe and the Taylor Museum in Colorado Springs. Many pieces have undoubtedly been destroyed or damaged and discarded during the intervening fifty years. Surely some of them remain undiscovered in private collections in New Mexico.

There are few major public collections of New Mexican tinwork. The largest group (more than 150 pieces), housed at the Museum of International Folk Art in Santa Fe, includes collections of nineteenth-century, Revival period, and contemporary works. Another significant collection is in the Taylor Museum of the Colorado Springs Fine Arts Center. With its acquisition of the Ortíz y Pino family pieces, El Rancho de las Golondrinas in La Cienega, New Mexico, became a major source for the study of New Mexican tinwork. The Colorado Historical Society owns a substantial number of tin pieces distributed among four different cultural sites: the Baca House in Trinidad, El Pueblo Museum in Pueblo, Fort Garland, and the society's museum in Denver. In Trinidad, the A.R. Mitchell Museum and Gallery has tinwork on exhibit in a morada reconstruction. The Southwest Museum in Los Angeles displays a few pieces from their collection in the Casa Adobe. Both the Kit Carson House and the Millicent Rogers Museum in Taos have modest collections of tinwork. Other museums with small collections are credited in the photo captions throughout the book.

NEW MEXICAN TINWORK, 1840–1940

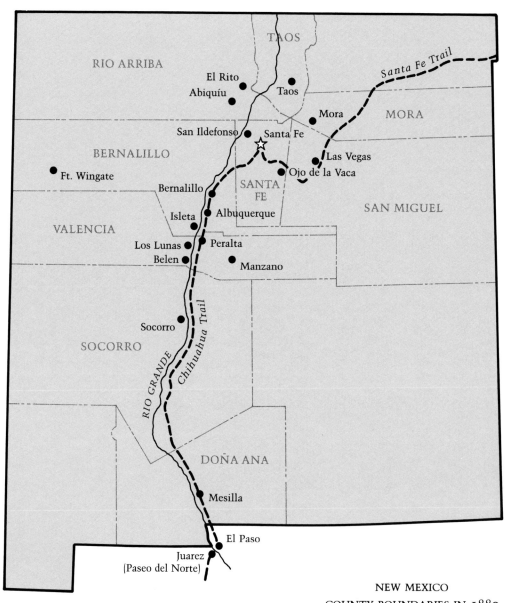

TAOS

RIO ARRIBA

El Rito

Abiquíu

Taos

Mora

MORA

Santa Fe Trail

San Ildefonso

Santa Fe

BERNALILLO

Ft. Wingate

Las Vegas

Ojo de la Vaca

SANTA FE

SAN MIGUEL

Bernalillo

Isleta

Albuquerque

VALENCIA

Los Lunas

Peralta

Belen

Manzano

RIO GRANDE

Chihuahua Trail

Socorro

SOCORRO

DOÑA ANA

Mesilla

El Paso

Juarez
(Paseo del Norte)

NEW MEXICO
COUNTY BOUNDARIES IN 1880

I

THE HISTORY OF SPANISH NEW MEXICAN TINWORK

*O*ne of the earliest references to the use of tin objects in New Mexico is found in the 1776 Domínguez inventory of the missions of New Mexico. Domínguez reported more than twenty objects made of tinplate, including numerous processional crosses, several boxes for altar breads and holy oils, a crown for a statue of a saint, one lamp, and a reliquary for "wax saints."[1] These were undoubtedly imported from Mexico or Spain and would not represent New Mexican tinsmithing. None of these objects seems to have survived into the twentieth century.

In many ways the development and decline of tinwork mirrors political and economic changes in nineteenth-century New Mexico. Although tin objects reached New Mexico by the early eighteenth century, Hispanic tinwork did not begin to develop as a craft there until after the materials became available about 1840. Some tin (or more likely tin-framed mirrors) and retablos painted on tin sheets may have been shipped north from Mexico over the Chihuahua Trail in the 1820s and 1830s. It is possible that these frames provided the prototypes for local craftsmen. The Santa Fe Trail was opened to trade from the United States in the late fall of 1821, but very little tin in any form was brought to New Mexico before the American occupation. Despite popular notions to the contrary, tin was scarce throughout the West during these years. From Utah in 1849, Brigham Young "counseled the immigrating saints to bring with them 'sheet tin of the best quality,'" saying tin was "better here than gold."[2]

The earliest reported tinsmith in New Mexico was an American, "Roberto," working in Santa Fe in 1826.[3] Roberto may have been the American craftsman referred to by Antonio Barreiro in 1832 when he derided the local Spanish people for not developing the craft.[4] He probably produced small kitchenware, pails, and other containers and did some repair work. There was no demand for roofing or flues for wood stoves at the time. Pitched tin roofs and gutters were not used until after the U.S. Army occupation some twenty years later, and wood stoves did not reach New Mexico until the 1860s.

Trade on the Santa Fe Trail greatly expanded during the 1830s and early 1840s, from sixty-three wagons loaded with $100,000 worth of goods in 1833 to more than $400,000 in merchandise shipped in 1843.[5] With the expansion of trade from the United States, a variety of previously unavailable goods began to enter New Mexico, at first shipped in wooden chests or barrels. Clothing and bolts of cloth, tools and foodstuffs, particularly coffee, were brought in by the traders along with some of the materials used by the first tinsmiths—wallpaper and mirrors. The two primary materials—tin salvaged from tin cans and window glass—were rarely available during these early years. The use of tin cans was not common or accepted by the public in the United

States until after 1840. Window glass began to be commonly imported after the American occupation in 1846.

The first specific reference to tinwork made in New Mexico appears in a description of the Military Chapel (La Castrense) on the plaza in Santa Fe. A traveler, Alfred Waugh, described the interior of the church shortly before the arrival of American troops in the summer of 1846: ". . . it was lighted by a rude chandelier pendant from the center of the ceiling and by tin sconces on the walls."[6] The chandelier may have been made of wood, but it is possible that it, like the sconces, was tin. This existence of tinwork made prior to the occupation suggests that at least a limited amount of tinwork was made in New Mexico in the period 1840–46.

In 1846 the Army of the West, commanded by General Stephen Watts Kearny, occupied New Mexico and declared it a territory of the United States. A number of soldiers under his command made descriptions of the interiors of churches and homes of the period. Lieutenant Abert, possibly describing the Parroquia in Santa Fe, wrote in August 1846: "The interior of the Church was decorated with some fifty crosses, a great number of the most miserable paintings and wax figures, and looking glasses trimmed with pieces of tinsel."[7] These framed mirrors very likely had tin frames rather than wooden ones decorated with metallic cord. Several other accounts from soldiers in the Army of the West referred to a local mid-nineteenth-century custom of covering the walls of both homes and churches with countless mirrors.[8] Mirrors augmented the dim light from the small windows and were a curiosity on the frontier. Surprisingly, very few early mirrors have survived in either wood or tin frames.

After the occupation of New Mexico, trade with the United States expanded rapidly. By 1847, the range of goods available in New Mexico included groceries and dry goods of all kinds, glassware, Queensware china, lead, gunpowder, soap, stationery, and tinware. A Santa Fe hotel advertised "oysters and sardines always on hand and cooked if desired."[9] These were some of the first products available in cans. A large portion of the goods were routed to the army garrisoned in Santa Fe. Small quantities of window glass for officers' quarters and large tin containers of lard and lamp oil were also imported over the Santa Fe Trail as quartermasters' supplies. The availability of materials and the demand for new tin products, particularly candle sconces, contributed to the growth of the craft in the late 1840s.[10] By 1850, four tinsmiths were listed in the City of Santa Fe census returns.[11]

In 1850, Pope Pius IX appointed Father Jean Baptiste Lamy as bishop of the new vicariate (later an archdiocese) of New Mexico. The French-born Bishop Lamy and his vicar general, the Reverend Joseph Machebeuf, arrived in Santa Fe from Cincinnati in the summer of 1851.[12] Lamy was determined to fill the mostly vacant missions and to bring modern church doctrine to the people of New Mexico. Upon his arrival he put Machebeuf in charge of restoring the Military Chapel for his use. W.W.H. Davis, describing the interior in the mid-1850s after the restoration, wrote: "A tin chandelier is suspended over the center of the cross [of the floor plan], and engravings of saints are seen in various parts of the house."[13] Shortly after his arrival, Lamy began to recruit priests from France to man the parish churches. By 1859, eighteen parishes had been established, staffed by at least ten French-born priests.[14] Some of the French priests, particularly those with funds available from their families, distributed devotional lithographs of saints to their parishioners.[15] They encouraged the use of the new printed images and discouraged the continued veneration of the old hand-painted retablos. The prints provided the greatest impetus for the development of tinsmithing in New Mexico.

European and American lithographs were also sold by European-born merchants in Santa Fe and Albuquerque during these years.[16] The prints (primarily lithographs) created a demand for frames made by tinsmiths. The simultaneous availability of large five-gallon lard cans, inexpensive religious prints, window glass, and wallpaper caused tinwork to flourish in New Mexico by the late 1850s.

By 1860 New Mexico had become a frontier territory of the United States rather than a northern outpost of New Spain, and the American occupation brought stability and some economic prosperity to the territory. The period from 1860 to 1890 was the time of greatest production for nineteenth-century New Mexican tinsmiths. Thousands of frames, sconces, nichos, crosses, and boxes were made in this period. During the late 1860s and the 1870s, the eastern railhead moved westward from Missouri. Closer shipping points for the Santa Fe Trail were created first at Independence, Missouri, then at Kansas City, and later at Leavenworth and Topeka, Kansas. During the winter months, wholesalers at these railheads advertised in New Mexican newspapers for the summer shipping season. By 1876, more than $2 million worth of goods were shipped

1.1

(see Figure 1.1)

annually to New Mexico on the trail.[17]

In 1879 the railroad reached Las Vegas, connecting New Mexico with the eastern United States. By the winter of 1880, the railroad had reached Albuquerque and pushed as far south as San Marcial. This allowed even greater quantities of goods and more fragile items to be shipped easily. As a result of railroad shipping, larger panes of glass became available to tinsmiths. Simultaneously, large-scale chromolithographs became popular in the United States and were distributed by merchants in New Mexico. Because the railroad provided safer transportation for larger pieces of glass, tinsmiths were able to make larger frames and nichos after 1880. The reduced costs of shipping allowed frivolous items

FIGURE 1.1
Detail of frame containing promotional calendar illustration. Rio Arriba Workshop, circa 1890–95, 10¹/₂ × 8¹/₂. Private collection, Santa Fe. (All dimensions are in inches.)

such as advertising cards and calendars to be given away to customers by the merchants. After 1880, these began to appear as graphics in tinwork frames and trinket boxes (see Figure 1.1).

The earliest known photograph in which tinwork appears was made about 1872. An interior view of the south chapel of the Santa Cruz church shows six Federal-style tin frames with lunettes and square corner bosses.[18] Other photographs of church interiors from the 1870s show numerous small tin candle sconces and a few frames hanging along side walls.

After 1890 the modernization of the previous forty years adversely affected the role of tinwork. Gas and coal oil lighting became commonplace. Candles, and consequently tin sconces, were no longer needed, while commercial picture frames replaced handmade tin frames.[19] Although three-dimensional bultos continued to be produced after the decline of retablos, by 1890 the demand for nichos had also ceased. The production of tinwork rapidly declined from 1890 to 1900; by the turn of the century, only a few tinsmiths continued to practice the craft.[20] Describing the mission at Isleta Pueblo in 1915, former Governor L. Bradford Prince wrote: "On the walls around, are several crucifixes of various ages and styles . . . and a multitude of mirrors and painted lithographs of saints, in the embossed tin frames which were so general in New Mexico until recent years."[21] The early period of tinsmithing in New Mexico had come to an end.

2

TOOLS, MATERIALS, AND PROCESSES

Some basic knowledge of the tools used by Spanish tinsmiths in New Mexico is helpful in recognizing the work of the major tinsmiths or workshops. A description of the wide range of materials, too, is useful for determining particular styles and dating individual pieces. The decorative processes used by the tinsmiths are part of a long tradition of surface embellishment developed in Spain or Mexico, while the joinery techniques differ little from traditional nineteenth-century tinsmithing methods employed in the eastern United States.

Tools

A limited number of tools were needed to fashion tinwork in nineteenth-century New Mexico: tin snips or large shears for cutting, a soldering iron for soldering the joints, and a source of heat for the irons. The tinsmith would also have had a set of steel punches for cutting and decorating the surfaces of tin sheets, and possibly a compass for laying out patterns. It is doubtful that the craftsmen producing decorative tinwork in New Mexico owned

2.1

FIGURE 2.1
*Selection of
tinsmithing tools and
materials, circa
1900–40. Private
collection, Santa Fe.*

were aware of beading machines and may have been imitating them in some of the scored details of frames.

Tin snips or readily available hand-forged shears could easily cut needed shapes from thin sheets of tin. Soldering irons, originally and correctly called soldering coppers, were made with a thick pointed bar of copper at the end and an iron shaft fitted with a wooden handle. The thick end retained the heat, and the solder adhered well to the copper. Nineteenth-century tinsmiths had small rudimentary forges for heating the coppers in their workshops, burning coal (if available) or charcoal. Blowtorches, which became available as sources of heat toward the turn of the century, were not used for soldering, but for heating soldering irons. The large flame of a blowtorch is too general to control the flow of solder and would discolor most of the tin surface.

All tinsmiths kept a variety of iron or steel punches and stamps to decorate tin surfaces in a stamping or embossing process. The punches were filed to shape and then hardened and tempered by the tinsmith or a blacksmith. The stamps were made from iron or steel bolts, rods, chisels, and other salvaged items. Some of the larger stamp images required that a blacksmith upset, or thicken, the end of a small rod, and long curved stamps had to be forged wider than the original stock on an anvil. Very few stamps were direct copies of leather-working tools as has been suggested by other writers. The small-scale details filed into the surface of these stamps are subtle and do not transfer well to tin. The dies used by tinsmiths were larger, with coarser detail.[4]

any more sophisticated equipment than this before the twentieth century. A Louis Prang chromolithograph from 1874 illustrates a typical eastern American tinshop of the period.[1] Visible are large iron stakes for forming and a combination rolling machine used for crimping stovepipes and embossing beading details. A similar establishment, the Pioneer Tinshop owned by A.G. Irvine and W.A. McKenzie, was in operation in Santa Fe by 1874, manufacturing "copper, tin and sheet iron ware."[2] In 1880, the Pioneer Tinshop produced more than $10,000 worth of goods.[3] We can assume that a tinshop of this scale would have owned the same production equipment as its counterpart in the East. There is some evidence that Hispanic tinsmiths

Materials

The primary materials used to produce New Mexican tinwork were tin, glass, paint, and wallpaper; the secondary materials were solder and flux. Other components were the engravings or lithographs that were included inside a frame or nicho, a discussion of which is included in Chapter 3.

Tin, or more accurately tinplate, was sheet iron coated with a thin layer of pure tin. This coating is safe to use in contact with foods, and tin plate was the material used to construct tin cans.[5] The invention of the Bessemer Process for making steel, circa 1850, and the improved open-hearth method, circa 1875, caused virtually all tinplate used for can making to become tin-on-steel by 1880. All nineteenth-century New Mexican tinwork (with a few rare exceptions) was made from salvaged cans.

Until circa 1893 New Mexican tinwork was constructed from British tin salvaged from American-made cans.[6] In 1891 the United States passed the McKinley Tariff Act, placing a high duty on imported tinplate. Following this act the American tinplate industry grew rapidly, until by the beginning of the twentieth century it had replaced England as the major producer of tinplate.

Tinplate came in a limited number of sizes of rather small sheets, designated by thickness, size, and quality of the tin coating. The two most common sizes were ten inches by fourteen inches and fourteen inches by twenty inches. The thickness was designated C, or common (about 29 gauge), to 1XXXX, the thickest (24 gauge). A tinman's manual for 1861 calls for two sheets of DX tin (12 1/2 inch by 17 inch of 26 gauge) to produce the body of an oil canister holding five gallons.[7] This size of can was the most common one used by New Mexican tinsmiths in the nineteenth century.

Another sheet material used in New Mexican tinwork was terneplate. Terne, thin iron or steel sheet coated with a lead-tin alloy, was used primarily for roofing, since the lead content of the coating was poisonous in contact with food and was not used to make cans. Terneplate has been observed on only a few pieces of the late nineteenth-century work of the Valencia Red and Green Tinsmith. Terne, darker than tinplate, was used almost exclusively during the Revival period of the 1920s and 1930s because it had the appearance of old oxidized tinplate. New tinplate is shiny and bright, but oxidation on older pieces is a slightly darker gray color, without gloss. Terneplate, still manufactured in West Virginia, is used by contemporary New Mexican tinsmiths.

Solders, the metal alloys used to join metal parts, were most commonly made from equal parts of lead and tin. Lower-temperature melting solder was made with two parts tin to one part lead.[8] Bar lead, essential on the frontier for making bullets, was commonly shipped over the Santa Fe Trail from the 1820s until 1880. The same suppliers in the East shipped solder as well.[9]

Flux, a solution that removes the surface oxidation and promotes the flow of the solder, also keeps the surface clean during the soldering process. The most common flux for lead (or soft) soldering is made from pine rosin mixed with a little oil. The oil makes the rosin less sticky and forms a liquid that can be brushed on the joints. The residue of rosin flux hardens into a deep amber crust

2.2

2.3

FIGURE 2.2
Sconce containing mirror. Unidentified workshop, circa 1870, 11⁵/₈ × 11³/₈. Collections of the Museum of International Folk Art, a unit of the Museum of New Mexico, Santa Fe.

FIGURE 2.3
Frame with hand-drawn "Santo Niño de Atocha." Rio Arriba Workshop, circa 1880, 6 × 5¹/₂. Collection of Ford Ruthling, Santa Fe.

around the joint and remains unless it is removed with solvent.

In her various writings about New Mexican tin-work, E. Boyd often refers to early tinwork that was held together with pine rosin instead of solder.[10] Her suggestion that this method of construction was a unique characteristic of very early tinwork in New Mexico has been accepted and repeated by other writers. An extremely close examination of a candle sconce, described and illustrated by Boyd as an example of a piece assembled with rosin (Figure 2.2), however, shows that in fact the piece is soldered at all the joints and that the soldering is original. The "pine rosin" is actually soldering flux that was never cleaned away by the tinsmith.

Old original glass seen in early tin pieces from New Mexico often contains irregularities such as inclusions, air bubbles, and press marks. The thickness of the glass also varies from sheet to sheet and from one edge of the glass to the other, with most original glass from ¹/₁₆ to ³/₃₂ of an inch in thickness. This is thinner than present-day single-strength glass. Mirror glass of the nineteenth century is also thinner than that of current mirrors.[11]

Several kinds of paint were used in conjunction with tinwork. Watercolor paint on paper was used under glass panes in a variety of ways. Decorative floral patterns are not uncommon, and numerous small watercolor paintings of Santo Niño de Atocha also can be found (see Figure 2.3). These small watercolors are all the same size, with the figure portrayed in the same position. All are pen-and-ink outlines filled in with watercolor, and the similar lettering on the pictures suggests that they are the products of a classroom exercise. They are probably from the Loretto Academy in Santa Fe, which was founded by Bishop Lamy and Sister Mary Magdalen in 1852.

Most of the back-painted glass panels were decorated with oil-based paints. Oil paints were brought into New Mexico over the Santa Fe Trail as early as 1847 in boxes of dry pigments with cans of linseed oil, varnishes, and turpentine.[12] The pigments could be mixed easily into a variety of paint colors. Unlike water-based paints (tempera and watercolor), oil paints will adhere to glass. The idea, espoused by some writers, that tempera paint was applied to paper which adhered to the glass as it dried, is quite unlikely. The major proponent of this style of painting, the Rio Arriba Painted Workshop, applied oil paints directly to the glass and, after it dried, backed the painting with paper. Some of the work of the Valencia Red and Green Tinsmith utilized oil paints that were extended considerably with linseed oil and applied to the surfaces of the tin, producing a very thick but transparent paint. Many of these pieces have suffered extensive paint loss owing to the brittleness of this paint layer and the flexing of the thin tin sheets during handling.

Processes

The primary processes in constructing and decorating New Mexican tinwork were cutting, piercing, stamping, embossing, scoring, and soldering. A secondary method of joining called tab construction was also used.

Cutting was usually accomplished with tin snips or large shears. Regularly scalloped edges of pieces were cut with a sharpened curved punch driven through the tin. Piercing was done by chiseling an opening large enough to insert shears in order to finish the cut.

There are a number of ways to develop an enriched surface on tinplate, including stamping and embossing. When stamping, the tin is hammered with a shaped steel punch from the front of the piece, leaving an image of the die in the surface. Dots and crescent shapes are the most common stamps, although serrated or notched straight-line punches and full circles were also used. All of the workshops created their own stamp patterns and combinations of stamp patterns which were unique to their workshops. A detailed description of these stamps can be found in Chapter 5.

Embossing is similar to stamping except that the tin is punched from the back and the image on the face of the tin is less distinct. Pieces are often worked with both techniques, using both sides of the tin sheet. The combining of both stamping and embossing creates the illusion of depth in the reflective surface of the tin.

Repoussé, a technique related to embossing in which designs are pushed out from the back in relief, was not often used in New Mexico. Sometimes a combination of embossing and stamping had the effect of creating a raised image, like one developed with the repeated hammer blows used in repoussé. The five- and six-pointed stars typically used in the lunettes of the Rio Arriba Workshop are the best example of near-repoussé but were created by scoring both sides of the tin sheet (see Chapter 5). The birds shown on the nicho lunette in Figure 5.18 are an example of actual repoussé raised in relief from the back of the piece. The outline of the image was punched from the front in a simple single-dot pattern.

Scoring, the process of drawing or stamping a straight line on the tin piece, was accomplished with a long, straight chisel stamp or with a pointed tool held against a straight edge and drawn along the line. This technique was common in the work of both the Santa Fe Federal and the Rio Arriba workshops.

Soldering is accomplished by fitting two tin parts closely together and fluxing the surfaces to be soldered. The soldering iron is heated, the solder is melted onto it, and the soldering iron is run along the joint, transferring solder to the tin. As it cools, the joint becomes solid. A variation for soldering blind joints is a technique called sweat soldering. In this method, the solder is applied with the soldering iron to both surfaces to be joined, and the pieces are then held together and reheated with a clean soldering iron. The heat, transferred through the tin, fuses both soldered surfaces.

Joinery techniques without solder are not common in New Mexican tinwork. The process of tab construction is simple; thin slits are chiseled in the tin, and a rectangular tab of tin from one part is slipped through and bent over to hold the piece in place. L-shaped wall sconces (described in Chapter 4) were commonly made with tab construction. The few examples of tabbed frames are later pieces that may have been made by amateurs. The concentric rosettes seen in the work of José María Apodaca (see Chapter 5) are joined with narrow strips of tin that pass through the layers and are folded over on both sides to lock the assemblage together. The tinsmith probably used this method to avoid scorching the painted surface of the tin salvaged from multicolored cans. Another joinery method that does not include solder—riveting—was not used by New Mexican tinsmiths.

Although the methods of fabricating New Mexican tinwork are typical of American tinsmithing of the period, the materials and surface decoration are unique to Spanish New Mexico. The almost exclusive use of salvaged cans and scraps of wallpaper and glass show how valued such materials were on the frontier and demonstrate the inventive skill of the New Mexican tinsmiths.

3
DATING NEW MEXICAN TINWORK

*D*ating New Mexican tinwork is problematic. Tin pieces may be dated from several sources of information—design styles of the pieces; wallpaper; can labels or embossings; European, Mexican, or American prints; and a few advertising cards all offer some clues toward determining the age of a piece. It must be pointed out that tin pieces were made with datable materials salvaged after those materials had served their original purpose. As a result, a given work cannot be dated to an exact year but only as coming after the most recent datable material used in the piece.

Since the tinwork in New Mexico was produced by such a small number of workshops, it is possible to assign working dates to these workshops from only a few bits of specific information. We synthesized dating using information from all possible sources—can labels from one frame, prints from other frames, wallpaper in another, and, if possible, patent dates. These assigned dates are reasonably accurate, based on information that is currently available.

Stylistic determinations of age are often the least accurate way to date an object, particularly one from

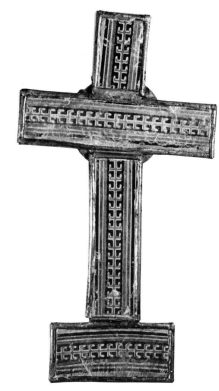

3.1

a folk tradition. The Santa Fe Trail trade and rail-road shipping brought the styles of the eastern United States randomly, in parts and pieces. In New Mexico in the last half of the nineteenth century, homes were not designed by architects and other arbiters of taste, but were fitted out as best the owners could afford from the limited supplies of local merchants. American Federal, Empire, Gothic Revival, and Victorian styles all overlapped and commingled during this period. The time lag between provincial design in New Mexico and the East, the extent of the exposure to outside design, and the understand-ing of styles by individual tinsmiths all make most generalities of stylistic dating improbable. Crude forms and rough craftsmanship do not connote an earlier date. We agree with E. Boyd's theory that the earliest pieces made in New Mexico reflect a clear understanding of the crisp lines and architectural forms of American Federal period frames as well as a knowledge of neoclassic details of pieces that may have come up the Chihuahua Trail from Mexico. Federal-style frames of the 1850s are often well crafted, while work from the turn of the century can be quite crude. The level of craftsmanship was determined by the skill of the tinsmith, not the period of production.

The designs of wallpaper fragments incorporated into tin frames or nichos are another source of information for dating, though only very general observations can be made. We know that wallpapers were introduced into New Mexico soon after the Santa Fe Trail was opened. An 1835 shipping manifest of the traders Langham and Boggs has an entry for a bundle of "13 reams of half flowered wallpaper."[1] An "invoice of . . . goods bought in St. Louis" by M.O. Yrisarri in 1853 includes "30 pieces of paper for the walls" and "1 piece of paper for borders."[2] While wallpaper was available even before tinsmithing developed, the greatest amount was used in late-Victorian brick homes built in Las Vegas, Albuquerque, and Santa Fe after 1880. The styles of wallpaper seen in New Mexican tinwork include pre-, early, and late-Victorian; William Morris; and Arts and Crafts period designs. The Victorian period (circa 1840–1900) spans virtually the entire period of early tinworking in New Mexico. William Morris had introduced his wallpaper designs in England by

1875. The more stylized patterns of the Arts and Crafts period were not common in the West before 1890. The turn-of-the-century pieces of José María Apodaca tend to contain more late-Victorian and Arts and Crafts papers than does the work of other tinsmiths.

The cross seen in Figure 3.1 may be one of the earliest pieces of tinwork extant. Mica sheets instead of glass are used to protect the pre-Victorian border paper. It is unfortunate that this piece does not have any tin details that might link it with other pieces. The wallpaper is a very early style, and the absence of glass suggests a date prior to 1850. At the time of the American occupation in 1846 very few buildings, with the exception of the Governor's Palace and a few large homes, had window glass. With troops permanently stationed in New Mexico, the army brought in glass to use in building officers' quarters. The trade generated by the needs of the army greatly increased shipping over the Santa Fe Trail, and a wider variety of goods began to come into the new territory.

Some of the new goods shipped over the trail came in tin containers. Francisco Sandoval, who started as a tinsmith in Santa Fe in the 1880s, remembered paying two cents a can for discarded army lamp oil and lard cans.[3] Early tinsmiths generally salvaged large, square five-gallon (or fifty-pound) lard cans about thirteen inches tall. Smaller cans for oysters, coffee, and tobacco were also used. Lard cans with markings from Kansas City, Chicago, Denver, and St. Louis have been found in New Mexico, as well as oyster cans from Baltimore and sardine cans from Maine.

The earliest complete patent markings we have

3.2

found stamped and embossed on New Mexican tinwork are August 22, 1862; June 23, 1863; June 28, 1864; and October 15, 1878. The last two dates are hand-stamped on the can, while the others are embossed with a drop press. All of the dates refer to patents for improvements of the tin can.[4]

The earliest method of labeling commercial tin cans was to emboss the label directly on the can or solder a separate embossed tin piece to the can. The most common embossed labels seen on New Mexican tin pieces are Fairbank and Co., St. Louis and Chicago, Pure Refined Family Lard; A. Booths Oysters, Baltimore; Thomas J. Biggers, Kansas City; and Plankinton and Armour, Kansas City. John Plankinton was a partner of P.D. Armour in both the Milwaukee and the Kansas City packing operations (founded in 1870). By 1880, the Kansas City plant

FIGURE 3.2
Detail of frame showing hand-stamped patent date. Possibly Santa Fe Federal Workshop, circa 1865. Harwood Foundation, Taos.

3.3

FIGURE 3.3
*Detail of frame,
embossed "A Booth's
Oysters." José María
Apodaca, circa 1900.
Private collection,
Santa Fe.*

metal and for ways to roughen the tin to accept the ink. The lithographed labels, which were one color—generally black—overprinted on a colored base coat, were common in the 1880s and 1890s and were produced well into the twentieth century. Chromolithography, or true color lithography, was not developed for printing can labels until the 1890s.[7] The White Cloud Compound label seen in Plate 1 is an example of two-color lithography with black-and-white printing on a solid red ground. Thus, any lithographed label dates at least post-1875 and usually after 1880.

Lithographed labels sometimes offer additional information for dating tin pieces. A label may include the name of the company that produced the can as well as the name of the product and the canning company. The lithographed labels on tinwork in New Mexico are often in remarkable condition, showing no signs of fading, weathering, or abrasion. It is probable that cans were salvaged almost immediately after use and made into tin items. Two events have a bearing on dating information on labels. First, the American Can Company was founded by three partners in 1901. They created a monopoly by buying as many as sixty regional can companies in the same year; as a result, almost all labels from small local can makers date before 1901. Second, in 1906 the first Food and Drug Administration Act was passed, requiring more honest labeling of foodstuffs, so that can labels after that date usually include a reference to the act.

A small rectangular frame in the collection of the Colorado Historical Society in Denver (#H3759, not illustrated here) has portions of a label on the back that serve to demonstrate how such labeling

was producing more than 700,000 cans annually, and by 1885 more than four million. The company is still in operation today as the Armour Company. W.W. Tate of Santa Fe advertised in 1882 that he was "the agent for A. Booth's celebrated oval brand of fresh oysters" (see Figure 3.3).[5] Thomas J. Biggers's company was founded in 1867 to serve the English and Irish markets for canned beef and lard but did not survive the recession of the late 1870s.[6] These labels are quite helpful in dating New Mexican tinwork.

Hand-stenciled images were used for tin-can labels in the 1850s and 1860s, but by far the most common process for labeling cans in the late nineteenth century was lithography. First developed in the 1870s, lithography on metal was initially hampered by a need for inks that would adhere to the

can help date a piece. This frame by the Rio Arriba Workshop is similar to the one illustrated in Figure 5.16. The label (reconstructed from additional labels) reads, "Colorado Packing and Provision Company. Denver, Colorado." In very small letters the label also reads, "The Eaton-Ritchell Co's." Portions of this black-on-yellow, one-color lithographed label are commonly seen on New Mexican tinwork. Colorado Packing and Provision Company, founded in Denver in 1893, was in business for many years. This information provides a beginning date in determining the age of the frame. The Eaton-Ritchell Company was also founded in 1893, by C.B. Eaton and E.C. Ritchell. Eaton started as a tinner as early as 1885 and formed several can companies in Denver with different partners, hiring Mr. Ritchell as a "japanner" with his Denver Stamping Company. Eaton-Ritchell became the Eaton-Ritchell Manufacturing Company in 1900, and by 1901 Ritchell had become the manager of the American Can Company in Denver.[8] From this information, then, we know that the can used in this frame was manufactured between 1893 and 1899 and the frame was produced during the period from 1895 to 1900. It should be noted that a similar frame by the same maker contains part of a calendar illustration marked "Compliments of W.N. Emmert. . . ." (Figure 1.1). Emmert was a grocer on the plaza in Santa Fe from 1889 to 1891, which suggests that this frame was made circa 1890–95.[9]

The devotional prints on paper that are commonly seen in tin frames and occasionally in nichos offer the greatest number of possibilities for determining the dates of works. The printmaking techniques used include those of woodcuts, engravings, lithographs, chromolithographs, and oleographs. The prints originated in Mexico, France, Germany, Switzerland, and the United States. While European and Mexican prints were not uncommon in New Mexico in the eighteenth century, their fragility and lack of protective glass caused them to disappear before tin and glass frames became available. The earliest prints found in tin frames are small, coarsely delineated Mexican woodcuts that date from the second quarter of the nineteenth century. Wood-block engraving was used much longer because of its compatibility with letter-press printing for newspapers and periodicals. The eighteenth- and early nineteenth-century steel or copperplate engravings of European and Mexican origin generally predate lithography but are not commonly found in tin frames. By far the most prevalent prints found in New Mexican frames are lithographs. The lithographic process, invented in Germany shortly before 1800, quickly became a fast, inexpensive way to produce commercial prints. French lithographers were the first to mass-produce popular images for the religious market, and during the period from 1820 to 1850 these prints became a major export item, particularly to the United States.[10] The major French lithographers of the mid-nineteenth century were the Turgis family, LeMercier, and Boasse Lebel.[11]

Turgis lithographs are the most common French prints found in New Mexican tinwork. Addresses included with the printer's name and the title of the image are helpful in determining the relative age of the Turgis lithographs. From 1828 to 1853, the firm was located at 16 rue Saint-Jacques, Paris, and at 36 rue de Rome, Toulouse. The firm moved in 1853, and the prints are marked "Vve Turgis 10 rue Serpente" (Paris). After 1856, the address was 80 rue des Ecoles, Paris. The lithographs imported into New Mexico were probably outdated by the time they arrived, and some of the Turgis images remained in print for a number of years. Prints with addresses that date from 1828 to 1856 are found in some of the oldest New Mexican tin frames.[12]

American printmakers adopted the lithographic process, which was quite common by 1840. Two American printmakers whose prints appear in New Mexican tinwork are the famous Currier and Ives of New York and Kurz and Allison of Chicago. Nathaniel Currier began his career as an apprentice in 1828. He opened his own firm in 1835, and in 1857 was joined by his son-in-law, James Ives. The prints were one-color lithographs, primarily hand-colored. As with the Turgis images, the dates of the prints sometimes can be determined by the address if it appears on the print. (Often in New Mexico, both the name of the printer and the title of the print have been trimmed to allow the print to fit an existing frame. Occasionally the title has then been superimposed over the bottom of the remaining portion of the print.)[13] Prints made before 1857 carry the name N. Currier, and after that date, Currier and Ives. Kurz and Allison Art Studios of Chicago produced numerous black-and-white religious lithographs for the widely separated concentrations of ethnic groups. Their prints have been found in the German areas of Texas and Wisconsin as well as the Spanish areas of New Mexico. The firm, founded by Louis Kurz in the 1860s, was active throughout the last third of the nineteenth century.[14]

Among the European printmakers who opened showrooms in the United States was the Turgis family of Paris, which had a store at 300 Broadway, New York City, in 1855 and 1856. The famed Benziger Brothers of Switzerland sold chromolithographs from their store in Cincinnati and later in St. Louis. Another firm, Haasis and Lubrecht, was in business in New York from the mid-nineteenth century well into the 1880s.[15]

Chromolithographs were introduced in New Mexico a few years after black-and-white and hand-colored lithographs. The Santa Fe stationer E. Andrews first advertised "pictures, chromos, frames and mouldings" in 1872.[16] Oleographs (chromolithographs imitating oil paintings) did not become commonly available in New Mexico until after 1890.

The dating of New Mexican tinwork must be a cumulative process. As more tin pieces are discovered which can be dated to a specific time period, our understanding of the dates for individual workshops will grow and become more specific. There is still a great need for a published study concerning the production periods of the various types of European religious prints; more information about can labels from midwestern packing houses would also greatly help the dating process.

4

FUNCTIONAL TYPES OF SPANISH TINWORK

*T*he New Mexican production of tinwork primarily for religious purposes is unparalleled elsewhere in American folk arts. Two of the most common forms of tinwork found in New Mexico, frames and nichos, were made almost exclusively for devotional use. Many lighting devices, candle sconces, and chandeliers were made for churches or family chapels as well. Tinwork produced solely for secular purposes is quite rare, although the most utilitarian objects—cups and coffee pots—either have not survived or have not been recognized as New Mexican. A discussion of the various forms made by Spanish tinsmiths in New Mexico will show that the demand for tinwork was instigated primarily by the need to display and protect European devotional prints that were introduced into New Mexico about 1850. Although our survey is not totally comprehensive, an understanding of the percentage of each function (i.e., frames, nichos, sconces) represented in the survey will give an idea how prevalent or how scarce each type was in nineteenth-century New Mexico.

4.1

Frames

Frames were the most common form of tinwork made in New Mexico, representing two-thirds of all surviving pieces. They were used for mirrors, religious prints, greeting cards, and even advertisements. Most of these were made to present and protect a variety of European engravings and lithographs of Roman Catholic saints that supplanted the earlier paintings on wood by local artists. They range in size from as small as two inches—a print worn as a locket—to nearly three feet tall. Early references to frames by travelers suggest that most

FIGURE 4.1
Frame containing mirror. Santa Fe Federal Workshop, circa 1860, 10¹/₂ × 10. Collection of Mr. and Mrs. J. Paul Taylor, Mesilla.

of the early ones contained mirrors.[1] Lieutenant Abert reported in 1846 that a house near Cerrillos had calico covering the lower half of the walls ". . . and pictures and looking-glasses of all sorts and sizes the upper half."[2] Painted and gilt-framed mirrors and prints had reached New Mexico as early as the fourth quarter of the eighteenth century.[3] Most certainly the earliest tin frames were patterned after the gessoed and paint-grained or gilded frames in the Federal and Empire styles brought over the Santa Fe Trail from the East after 1821 and the neoclassic frames that came up the Chihuahua Trail from Mexico in the first half of the nineteenth century.

Late eighteenth-century Federal-style frames were characterized by the use of half-round column-framing members and square corner bosses. Most Federal pieces had elaborate broken-pediment crests or carved and gilded American Eagle motifs, while some had a simple half-round lunette mounted at the top. The sides were decorated with abstract pendant leaf forms or scrolls (see Figure 4.1). The nineteenth-century Empire frames were more severe and simple, utilizing a rectangular frame of wide ogee molding with mitered corners. The less expensive paint-grained Empire mirror frames shipped into New Mexico in the mid-nineteenth century were square with rounded corners and ovolo moldings. These, then, were some of the prototypes available to tinsmiths in New Mexico.

New Mexican tinsmiths invented fanciful versions of the available prototypes and, although restricted by the severe rectangular format of mirrors and prints, they invented an incredible variety of shapes. Pendant leaf-form additions became impor-

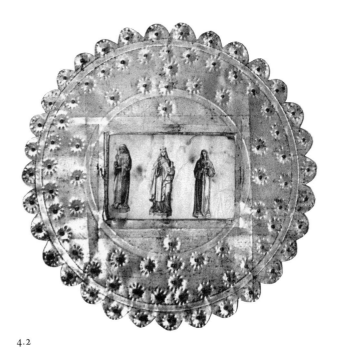

4.2

4.3

tant design elements (see Figure 5.7), and some makers even incorporated multiple leaf or scroll shapes at the lower edges of the frames. The lunette that had been in use as a crest on wooden retablos and New Mexican furniture of the late eighteenth and early nineteenth centuries continued to be a favorite form, appearing on the majority of tin frames. Half-round columns adapted from Federal frames were transformed into Spanish Solomonic columns which were commonplace on early altarpieces in New Mexico (see Figure 5.2). Simple low, triangular pediments, common on Empire period mirrors, were seldom used in New Mexico except as a stamped detail on a semicircular lunette. A few rare frames incorporated multiple attenuated pediments and the

trefoil finials reminiscent of the Gothic Revival style popular in the United States from 1830 to 1870.[4]

Most of the frames were square or rectangular shapes made from four strips of tin joined at the corners. The joints were covered with applied rosettes, square corner bosses, or scored and pleated fan shapes. Other common shapes were simple one-piece round frames made from the bottom panel of large cans, scalloped around the perimeter. The frames were pierced with a round or rectangular opening to receive a mirror, Holy Card, or devotional image (see Figure 4.2). Several makers designed frames that seem to be based on piecing the salvaged bits of scrap tin together into elaborate

FIGURE 4.2
Frame with religious catalog page. Rio Arriba Workshop, circa 1885, 11¹/₈-inch diameter. Private collection, Santa Fe.

FIGURE 4.3
Sconce. Valencia Red and Green Tinsmith, circa 1885, 13-inch diameter. Private collection, Santa Fe.

4.4

FIGURE 4.4
Frame with catalog page and wallpaper. José María Apodaca, circa 1895, 11 × 17. Taylor Museum, Colorado Springs, Colorado.

the back panel to hold a strip of torn cloth or cord for hanging. A frame occasionally had two loops on adjacent sides of the back, allowing it to be hung either horizontally or vertically. This suggests that the tinsmiths made the piece before a print was selected, or that the frame was designed to hold a mirror.

As design variations, several tinsmiths incorporated bits of wallpaper set under glass, surrounded by narrow tubes of tin soldered to the backing plate. These panels were then soldered around the perimeter of the main frame or image. The scarcity of glass in nineteenth-century New Mexico caused tinsmiths to sometimes piece together larger panels from small pieces of salvaged glass, occasionally covering the joints with scalloped strips of tin (Figure 4.4). This salvaging of both tin and glass is part of the unique construction of New Mexican tinwork. While cherished prints of saints were protected from household dust and smoke by tin and glass frames, reflections from the stamped tin provided a bright accent to the dark interiors of homes and churches in New Mexico.

Nichos

In the devout Spanish Catholic home, the *nicho* (or niche made to contain a holy image) was accorded a place of honor. Two purposes of the nicho were to house and protect important bultos and to serve as a sacred shrine for the family. In churches, nichos held images of the patron saint of the village as well as other revered saints. As a result of the importance placed on the nicho more tin nichos

scalloped, openwork forms with multiple rosettes and crescent shapes (see Figure 4.3). Still other shapes employed by the tinsmiths in New Mexico were complex octagonal frames and rectangles so modified with multiple lunettes or extended and enlarged pendant leaf forms that they must be considered eccentric.

Most frames had small strips of tin rolled into partial tubes and soldered onto the back of the frame to accept the glass and the tin backing panels. All the frames had some form of tin loop soldered to

have survived than have more commonplace tin objects such as wall sconces and candle holders. Early collectors were particularly intrigued with the elaborate sculptural qualities of nichos, and they collected them in some quantity. For these reasons, nichos are the second most common form of early tinwork to survive in New Mexico.

The earliest precedents for tin nichos were small wooden ones made throughout the first half of the nineteenth century by santeros. Wooden nichos were made either in the form of a small box—with or without a door—or as a framework of boards that defined the floor, ceiling, and corners of a box for the display of bultos.[5] While some nichos were made to hold bultos, tinsmiths in New Mexico expanded both the form and the function of these traditional religious objects by soldering small frames (containing devotional prints or Holy Cards) inside the nicho. Space was allowed in front of the picture for changing the decoration of paper or dried flowers and *milagros* (see Plate 6). One amazing example contains an early tintype photograph of La Conquistadora, the sculpture of the Virgin housed in the parish church in Santa Fe since the eighteenth century (Figure 5.28). These nichos were certainly not designed to hold bultos, but to serve as a family shrine that might be decorated. Nichos range in size from six-inch-tall boxes only two inches deep to extremely large forms more than three feet tall and at least a foot deep. Larger nichos were usually made for churches, where tall ceilings and larger spaces would accommodate or require more massive tinwork.

Tin nichos are in the form of three-dimensional rectangular boxes with hinged glass doors and side

4.5

panels framed by tin tubes. The back panel was made from a solid sheet of tin, often showing the original folds or seams of the tin container from which it was salvaged. Some tinsmiths stamped the backing plate with complex grid or diaper patterns and applied diagonally scored Solomonic half-round columns to the inside corners of the case. Most nichos had side panels soldered to the backing plate, shaped and decorated much like the side panels of frames. Wallpaper set under glass panels was frequently used as an additional side element, along

FIGURE 4.5
Nicho. Unidentified tinsmith, northern New Mexico, circa 1880, 28 × 19⅞ × 5, extensively stamped, embossed, and scored. Private collection, Santa Fe.

4.6

FIGURE 4.6
Sconce, one of a pair. Possibly Valencia Red and Green Tinsmith, circa 1875, 8 × 4 1/2 × 3. Collection of Shirley and Ward Alan Minge, Casa San Ysidro.

zoidal in cross-section, with tin side panels stamped to match the front and canted sides of the case (see Figure 5.13).

Nichos are the one tin form in New Mexico most often ornamented with cut and applied birds. The birds vary in shape from tiny demure chicks to menacing vultures. One can only speculate that these birds may represent the Holy Spirit in Christian iconography; they may merely be a secular design motif derived from sources such as the American federal eagle on commercial looking-glass frames, or the Hapsburg eagle. They appear on nichos as single or double images, and on frames with up to eight birds alternating with rosettes applied around the perimeter (see Figure 5.33).

While much of the design of the nicho came from the same American prototypes adapted to tin frames, the usage was purely Spanish Roman Catholic.

Sconces and Other Forms of Lighting

with pendant leaf forms or scroll-shaped appendages. The top of the nicho often featured a large lunette or a low pedimented crest (see Figure 4.5).

Nichos were equipped with two loops on the back connected with fabric strips or cord for hanging. A few have two curious large tin tubes soldered beneath the floor, probably to act as sockets for poles to hold the nicho aloft during religious processionals or to stabilize the large form on the altar of a church. Several tinsmiths in New Mexico developed a specialized shape of nicho that was trape-

Simple L-shaped sconces (*pantallas*) were very likely the earliest form of tinwork made in New Mexico. Constructed to hold a single candle, tin sconces did not burn as would those made of wood. They were the principal lighting fixtures in homes and churches in the nineteenth century. Other forms of lighting commonly made by the tinsmiths were frame sconces; small, easily portable candle holders; and lanterns (*faroles*). Less often, tinsmiths produced candlesticks (*candeleros*); candelabras; and, for churches and large ranchos, a few chandeliers (*arañas* [spiders]).[6] Few of these lighting fixtures survive today. Of the early tinwork in collections, less than 10 percent are lighting fixtures. Twentieth-century collectors may have ignored the simple sconces, or they may have been discarded by owners when electricity made their function obsolete.

The most common type of sconce was the simple projecting L-shaped type that was possible to fabricate without the use of solder. Referring to the church at Santa Cruz, New Mexico, in 1881, Lieutenant Bourke described the light produced by these sconces: ". . . tallow candles in tin sconces, affixed to the whitewashed walls lit up the nave and transept with a flicker that was [sic] in the language of poetry might be styled a 'dim religious light,' but in the plain, matter of fact language of everyday life, would be called dim only."[7] Numerous early photographs of New Mexican churches show these ubiquitous sconces lining both sides of sanctuaries. They were very common as late as the 1930s in the more remote areas of Hispanic New Mexico before the coming of the Rural Electric Power projects, and one mission church in northern New Mexico still retains a complete set of sixteen matching sconces.

L-shaped sconces were made from two pieces of tin—a rectangle piece bent out at the bottom to support the separate tubular candle socket (Figure 4.6). A hole at the top of the sconce allowed it to hang flat against the wall. Occasionally, the bottom shelf was folded to form a shallow box that would catch the candle drippings (Figure 4.7). Only rarely was a sconce made with a separate reflector attached behind the candle to augment light pro-

4.7

duced by the glow of the flame (Figure 4.8). The L-shaped sconces were usually very plain but sometimes were decorated with scored and embossed details.

More elaborate forms of the wall sconce were also made by New Mexican tinsmiths, often for churches. These sconces were constructed by adding a candle holder and *bobeche* (a saucer to catch wax drippings) to a shaped background plate, with a supporting bracket replacing the bent L of more simple sconces (see Plate 2). Intricately shaped, em-

FIGURE 4.7
Sconce, circa 1875,
10¹/₄ × 4¹/₂ × 2⁷/₈.
Collected in Taos
County. Collection of
Larry and Alyce
Frank.

4.8

4.9

FIGURE 4.8
*Sconce, circa 1900,
7 × 5 × 2, tab
construction,
terneplate. Kit Carson
Foundation, Taos.*

FIGURE 4.9
*Candle holders, set of
three. Possibly
Valencia Red and
Green Tinsmith, circa
1870, 2 × 5³/₄
diameter. Collection
of Shirley and Ward
Alan Minge, Casa
San Ysidro.*

bossed, stamped, and sometimes painted, such sconces present some of the most inventive shapes in New Mexican tinwork while still retaining the design elements typical of individual workshops.

Most of the tinsmiths in New Mexico produced a few wall sconces by attaching candle sockets and bobeches to the bottom corners of their frames. The candles may have been intended to illuminate the print or to light an area for devotions. Candle holders were added to the sides of nichos for similar reasons (see Figure 5.47). Occasionally, a candle holder might be placed inside the glass case of a nicho. Frame-style sconces as well as nichos incorporated methods of decoration identical to those utilized on standard frames and nichos: wallpaper panels, back-painted glass panes, painted tin, and stamping.

Another lighting form that must have been quite common—although few such pieces survive today—are small hand-held candle holders. Shaped like small saucers and occasionally decorated with stamping, they were used to carry light from room

4.10

4.11

to room (see Figure 4.9). Photographs show that they were often found surrounding the altar in the small churches and moradas of New Mexico.

Tin lanterns (faroles) were also produced by the tinsmiths. Patterned like tin lanterns of northern Europe and New England, they had a cylindrical body and conical top with a tin loop attached for hanging. The body was pierced with holes in complex patterns that allowed the light of a candle to shine through. This kind of lantern, developed as a more windproof light source, is commonly associated with outdoor Penitente processions (see Fig-

ure 4.10). Another style of lantern is a glass-paneled design that may have been adapted from eastern prototypes. This style, although rare in the nineteenth century, was common during the Revival period, when it was often electrified (see Figure 4.11).

The forms of lighting least frequently produced by nineteenth-century New Mexican tinsmiths were chandeliers, candelabras, and candlesticks, since they were much more complicated to make. Fewer than a dozen old chandeliers have been located. Their collection history indicates that they were made to

FIGURE 4.10
*Lantern. Unidentified
workshop, circa 1880,
14¹/₂ × 6 diameter.
Collected from
morada in Talpa,
New Mexico.
Collection of Larry
and Alyce Frank.*

FIGURE 4.11
*Lantern. Unidentified
tinsmith, northern
New Mexico, circa
1875, 13 × 6.
Collection of Mr. and
Mrs. Murdoch
Finlayson, Santa Fe.*

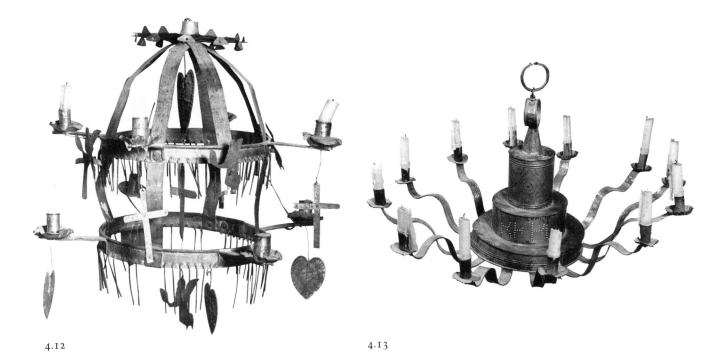

4.12   4.13

FIGURE 4.12
Chandelier (araña),
tin and copper.
Unidentified
tinsmith, circa 1850–
1900. Collected from
upper Arroyo Hondo
morada, Taos County.
Collection of Taylor
Museum, Colorado
Springs, Colorado.

FIGURE 4.13
Chandelier (araña).
Unidentified
tinsmith, circa 1860–
80, 15 high, 30
diameter. Spanish
Colonial Arts Society,
Inc., collection on
loan to the Museum
of International Folk
Art, Santa Fe.

be hung in churches, dance halls, and large residences. A tin chandelier is mentioned by Davis in
his description of the Military Chapel (La Castrense) on the plaza in Santa Fe in the mid-1850s.[8]

Chandeliers were constructed with one or more
circular bands of tin supporting extended arms which
terminated with the candle holders (see Figures 4.12,
4.13, and 5.56). There were usually six to twelve
candle holders, although an extremely complex
chandelier from the church of Nuestra Señora de
Belen in Belen, New Mexico (see Figure 4.14), has
more than forty candles in its three tiers. This chandelier is in the style of the *corona* or crown form.
The chandeliers usually were suspended from ropes

that allowed them to be raised or lowered to replace
the candles. Twentieth-century Revival tinsmiths
made numerous electrified chandeliers for residential and commercial buildings.

The candelabra in Figure 4.15 is the only known
early example of this form of lighting. The candlesticks in Figure 4.16 represent one of only two nineteenth-century pairs we have seen. The forms may
have been derived from brass candlesticks brought
to New Mexico by the clergy in the eighteenth and
nineteenth centuries. Candlesticks and candelabras were a common production item during the
Revival period of the 1930s, and they are still being
produced today.

4.14

Crosses

Considering the special place the cross has in the imagery of Spanish Catholic culture, it is surprising that tinwork crosses are not common, making up less than 5 percent of the tinsmiths' production. Although scarce, they were one of the earliest forms of tinwork made in New Mexico (see Figure 3.1). Tin crosses have been collected throughout the Rio Grande valley, from north of Taos to Mesilla, Doña Ana County, in southern New Mexico. They were generally made for use in homes or family chapels; churches were equipped with specialized crosses used in religious processions. (These will be discussed later in this chapter.)

Most of the principal workshops produced some style of cross, the pieces ranging in size from eight inches to almost three feet tall. All were in the form of the Latin cross, and they typically were made by inserting decorative wallpaper scraps between protective glass panes and a tin backing plate. The edges were bound with tin tubes. Virtually all of

FIGURE 4.14
Photograph of interior of Nuestra Señora de Belen, Belen, New Mexico, April 1894. Unidentified photographer (possibly Hans Becker). Courtesy of Dorothy D. and Austen Lovett, Belen.

4.15

4.16

FIGURE 4.15
*Candelabra.
Unidentified
workshop, circa 1860,
8 × 11 × 5. Formerly
in the collection of
Dr. Nolie Mumey,
collected in Taos. The
Denver Art Museum
(#1968-113), gift of
the May D & F
Company, 1968.*

FIGURE 4.16
*Pair of candlesticks.
Possibly Rio Arriba
Workshop, circa 1885,
28 × 4 × 4. Museum
of International Folk
Art, on loan to the
Palace of the
Governors, a unit of
the Museum of New
Mexico, Santa Fe.*

the crosses were intended to be displayed on a wall and were fitted with a tin loop on the back to hold a cord or strip of cloth for hanging. The ends of each of the arms of the cross had shaped tin finials that frequently were crescent shaped, reminiscent of the early Spanish convex iron hocking knives used during the Colonial period (see Figure 5.55).[9] Finials also took the form of fat reverse scrolls (Figure 4.17). The intersection of the arms was sometimes covered with a scored and embossed tin rosette, and adjacent areas between the arms (representing rays)

were filled with similarly decorated quarter-round brackets (Figure 4.18). A few tin crosses have diagonal braces connecting the arms of the cross, patterned after earlier wooden crosses made in New Mexico (see Figures 4.19 and 4.20).

While wallpaper under glass was the most common decorative material used in New Mexican tin crosses, other materials were also employed by the tinsmiths. Figure 4.17 shows a cross with paint applied directly to the back of the glass. Another cross, housed in the collection of the Southwest Museum

4.17

4.18

in Los Angeles, is mounted with pieces of early mirror glass. Figure 4.18 shows original oil-paint decoration both under the glass panels and on the surface of the tin. The cross in Figure 4.21 has similarly decorated tin and a unique form of decoration—cloth strips laid under the glass. The cross in Figure 5.92 has an early Mexican woodcut framed under glass in the bottom panel and combed paint in black and brown over a paper backing on the remainder of the four arms. Plate 3 shows yet another method of decoration used by the tinsmiths—

colored paper with pierced patterns overlaid on a paper of contrasting color. This cross is also painted, in part directly on the tin surface.

The crosses illustrated in Plate 3 and Figure 4.22, the only freestanding crosses known to exist, must have been designed to rest on an altar, in either a church or a family chapel. The cross in Figure 4.22, which reportedly was collected from the Santuario at Chimayó, is painted a deep blue that matches other pieces collected from the same source.[10] The hollow hexagonal base is embossed with rosettes,

FIGURE 4.17
Cross. Unidentified workshop, circa 1890, 14³/₄ × 10. Southwest Museum, Los Angeles, California.

FIGURE 4.18
Cross. Valencia Red and Green Tinsmith, circa 1885, 22³/₄ × 17, paint (possibly tempera) under glass. Collection of Ford Ruthling, Santa Fe.

4.19

4.20

and the tubular arms of the cross are capped with curious bullet-shaped finials.

Trinket Boxes

Tin boxes, called *baulitos* (little chests) or *cajitas* (small boxes), are rare in New Mexico. Fewer than forty are known to exist in public or private collections. Used for holding documents, makeup, jewelry, or other small personal items, they are often referred to by collectors as trinket boxes. Like lighting fixtures, boxes represent a type of tinwork made for secular use. The tin boxes resemble a much earlier form of European box known as a casket, traditionally used in Spain as a reliquary. Boxes, which were made in New Mexico by several of the workshops, have been collected in both the Rio Abajo and the Rio Arriba areas. They were also a staple of the tinsmiths' production during the Revival period.

Trinket boxes typically are rectangular forms

4.21

4.22

with a three-paneled hinged lid shaped much like the lids of hutch-top chests. The boxes were made from salvaged cans, with glass panes framed by tin strips over wallpaper fragments with a tin backing plate. Sometimes, though rarely, the boxes have shaped corner brackets at the bottom that function as feet or beveled moldings that form a base to the box. Decoration consists of stamped and embossed strips overlaid along the joints and similarly worked end panels on the lid (Figures 4.23–4.25). The boxes are equipped with a two-part hasp on the front which

serves as a handle for the lid.

In addition to Victorian and Arts and Crafts period wallpapers under glass, other decorative elements were used on trinket boxes. The box in Figure 5.32 shows the use of floral designs painted and combed in oil paints directly on the back of the glass panels. Plate 4 illustrates an elaborate box decorated with floral designs, with the owner's name in colorful watercolor paint on paper set under glass. Another example of decoration (not illustrated) can be seen on a box made from salvaged pieces of etched

FIGURE 4.21
*Cross. Circa 1885,
27 1/4 × 20,
polychrome tin,
fabric, glass. Private
collection, Santa Fe.*

FIGURE 4.22
*Altar cross.
Unidentified
tinsmith, circa 1875,
21 1/2 × 15 × 5,
polychrome tin,
originally attached to
tin staff collected
from the Santuario de
Chimayó. Taylor
Museum, Colorado
Springs, Colorado.*

FIGURE 4.23
Trinket box. José María Apodaca, circa 1900, 4³/₄ × 7¹/₂ × 4¹/₂, tin, glass, wallpaper. A.R. Mitchell Memorial Museum and Gallery, Trinidad, Colorado.

FIGURE 4.24
Trinket box. José María Apodaca, circa 1900, 9⁷/₈ × 8¹/₂ × 5³/₄, tin, glass, wallpaper. Collection of Shirley and Ward Alan Minge, Casa San Ysidro.

glass from a commercial entry door of the Victorian era. This piece surely dates after 1880, when such fragile and bulky objects could be safely transported to New Mexico by railroad.

A further example of inventive decoration used by tinsmiths is shown in Figure 4.26. This small box was made by Felipe Garcia of Santa Fe about 1900. It is constructed entirely from a single salvaged coffee can from a firm in St. Louis. The maker

used the deep blue of the can label as well as the golden lacquered surfaces of the can for decoration. Surfaces are further enriched by modest stamping and embossing. Mr. Garcia was a shoemaker by profession who did tinwork as a sideline. According to family tradition, he traded tinwork for pumpkins and other produce in the Ojo de la Vaca area of Santa Fe County in the fall. This piece has descended through the maker's family.

4.25

4.26

Other Uses of Tin

Traditionally, eastern tinsmiths fashioned functional items for the home. Manufactured kitchen utensils included cups, coffee pots, dippers, graters, cookie cutters, and baking pans. Wash boilers and buckets of all kinds were also commonly made. Tinsmiths built standing-seam roofs from tin plate, or more commonly terneplate, as well as gutters and downspouts. In contrast, the early Hispanic tinsmiths of New Mexico primarily made religious objects, though they also produced some items for secular use.

A variety of unusual religious objects made by tinsmiths has been discovered, including tin crowns for bultos. Often, these crowns were added to a bulto that had been made at an earlier date. The crown shown in Figure 4.27 appears to have been gessoed in place by its maker, José Raphael Aragón of Cordova, New Mexico.[11] This particular santero was working in northern New Mexico in the period from 1820 to 1862 and certainly would have been

FIGURE 4.25
Trinket box. José María Apodaca, circa 1900, 2⁷/₈ × 3⁵/₈ × 2³/₈, tin, glass, wallpaper. Collection of Shirley and Ward Alan Minge, Casa San Ysidro.

FIGURE 4.26
Trinket box. Felipe Garcia, circa 1900, 3¹/₂ × 6¹/₈ × 4, tin and glass. Collection of Senaida and Emilio Romero, Santa Fe.

4.27

4.28

FIGURE 4.27
Detail, bulto of San Miguel. José Raphael Aragón, circa 1850– 60, polychrome, wood, tin. Collection of Larry and Alyce Frank.

FIGURE 4.28
Tin crown. Unidentified bulto, nineteenth century, 4 inches high. Collection of Ford Ruthling, Santa Fe.

familiar with the use of tin during the latter part of his career. The piece illustrated has not been restored or altered from its original appearance in any way, which indicates that the tin is contemporary with the bulto. We consequently assume that it is one of the earliest datable tin pieces in New Mexico. The bulto of San Miguel carries his original sword and a set of tin scales. The crown shown in Figure 4.28 is more typical of those usually found on bultos. It has embossed and scored decoration

in addition to an applied tin Maltese cross at the top.

Figure 4.29 shows a unique use of tin in conjunction with a wooden piece made by a santero. Collected in Taos County, the carved, gessoed, and painted framework was intended to be fitted with a tinwork nicho. The bulto is attributed to the style of the Arroyo Hondo Painter, dated 1820–40, while the tinwork nicho was made by the Mora Octagonal Workshop. Examination of the paint wear on the

4.29

4.30

FIGURE 4.29
*Nicho on wooden
stand. Mora
Octagonal Workshop,
circa 1850, 22 × 11¹/₂
× 5¹/₂, tin, glass,
polychrome wood.
Collection of Ford
Ruthling, Santa Fe.*

FIGURE 4.30
*Host box. Rio Arriba
Workshop style, circa
1850–60, 2¹/₂ × 3⁷/₈,
reported to have
belonged to Padre
Martínez of Taos.
Collections of the
Museum of
International Folk
Art, a unit of the
Museum of New
Mexico, Santa Fe.*

base of the structure indicates that the existing ni-cho is very likely original. This nicho enclosure is the only complete example we have seen, although the Museum of International Folk Art owns two somewhat similar structures of unknown function which lack either a tin nicho or a seated bulto.

The small round box in Figure 4.30 is the only known surviving example of a priest's Host Box. It has long been in the collection of the Historical Society of New Mexico and (according to an old label on the bottom) was originally the property of Fr. Antonio José Martínez of Taos. If the label is

4.31

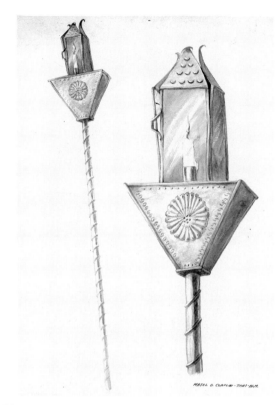

4.32

FIGURE 4.31
*Processional set.
Valencia Red and
Green Tinsmith, circa
1880, overall height
92, polychrome tin,
wood. Collected in
Manzano, New
Mexico. Collection of
Shirley and Ward
Alan Minge, Casa
San Ysidro.*

FIGURE 4.32
*Watercolor
illustration of pair of
processional torches
by Majel Claflin,
1937. Unidentified
tinsmith, circa 1870.
The set was collected
from a* morada *in
Mora, New Mexico.
Collection of* Index of
American Design,
*National Gallery of
Art, Washington,
D.C., #NM-ME-57.*

correct the piece would date circa 1850–60. The box's lid is decorated with a scored star much like those that appear on the lunettes of Santa Fe Federal and Rio Arriba Workshop nichos (see Figures 5.2 and 5.18). Scored details around the body are imitations of beaded lines produced by a combination rolling machine.

Another type of tinwork made for religious use is the three-part processional set made for churches. These sets consist of a cross on a stave and a pair of long processional torches that were displayed to the side of the altar or stored in the sacristy. They were used in processions, particularly for funerals or during Holy Week and saints' days. The pieces, about six feet tall, are made from tin attached to a wooden framework. The cross shown in Figure 4.31, collected from the church at Manzano, is one of the few surviving processional sets found in New Mexican collections. It is of hollow construction, and the long staff is covered with tin sheets salvaged from cans with "Pure Lard" still visibly embossed on the shaft. The set was made about 1875 by the

4.33

4.34

Valencia Red and Green Tinsmith (see Chapter 5).

Figure 4.32 shows a set of two processional torches made in the form of glass-enclosed lanterns (see Figure 4.11 for a lantern undoubtedly by the same maker). These pieces were collected from a Penitente morada at Mora, New Mexico, and were rendered in watercolor for the *Index of American Design* in 1937.[12]

The flower holder shown in Figure 4.33 is one of a pair from a private collection. They are part of a unique group that includes three more flower holders and two matching candle holders now in the Museum of International Folk Art. The set was made by José María Apodaca at the turn of the century, and probably was commissioned for a family chapel.

Although not strictly a religious piece, the small frame shown in Figure 4.34 was made to be worn as a locket or pendant. The surface of the tin remains bright and has been polished through wearing. It contains a tiny lithograph of Santo Niño and is one of only two located thus far in New Mexico,

FIGURE 4.33
Flower holder. José María Apodaca, circa 1900, 3-inch diameter. Private collection, Santa Fe.

FIGURE 4.34
Locket with lithograph of Santo Niño de Atocha. Late nineteenth century, $1^3/_8 \times 1^1/_8$, tin. Collection of Shirley and Ward Alan Minge, Casa San Ysidro.

4.35

4.36

FIGURE 4.35
Vase. Mid-nineteenth century, 9¹/₂ × 6¹/₂ diameter, tin, traces of pigment. Kit Carson Foundation, Taos.

FIGURE 4.36
Candle mold. Mid-nineteenth century, 12¹/₂ × 5 x 3¹/₄. Kit Carson Foundation, Taos.

though similar diminutive frames are not uncommon in Mexico.

Utilitarian objects made by New Mexican tinsmiths are not generally distinguishable from those made elsewhere in the country. Most buckets and tin cups have been worn out and discarded. The few pieces illustrated here are the only ones we have located. The vase shown in Figure 4.35 is made in the form of a typical (but incomplete) coffeepot similar to those found in Pennsylvania and New England. It is obviously made by a trained tinsmith,

as revealed by the rolled edge at the foot and the overlapping seam at the medial line. Another piece with Anglo influence is the candle mold in Figure 4.36. Spanish New Mexican candles were made by dipping; molds are associated with immigrants from the East.[13]

The only full-size furniture piece with tin details that has been located is illustrated in Figure 4.37. This chest may have been rebuilt from parts of an older piece, the original decorative parts assembled with wire nails. The lumber is milled pine

4.37

FIGURE 4.37
*Pine table with
applied tin
ornamentation, "Mr.
Lorenzo Martínez,"
"1872." Northern New
Mexico,* 30¼ × 38½
× 15. *Collection of
Peter Goodwin, Santa
Fe.*

4.38

FIGURE 4.38
*Two miniature chairs.
Unidentified
tinsmith, circa 1900,
left: 10¹/₄ × 5¹/₄ ×
4¹/₂, right: 10¹/₂ × 6
× 4¹/₂. Collection of
Sheila and Julian
Garcia, Albuquerque.*

trimmed with tin escutcheons behind the hand-made galvanized wire bail handles and decorative corner pieces. The tin was punched two pieces at a time, then attached to the wood with small wire staples. A decorative center piece has been lost, but the image of the stamping remains embossed into the wood surface, reading "Mr. Lorenzo Martinez" and dated 1872.

The unique pair of miniature chairs seen in Figure 4.38 was probably intended to be used as doll furniture. Dating from the early twentieth century, the chairs are patterned after late-Victorian factory-made armchairs that became available in New Mexico with the coming of the railroad in 1880. Although somewhat crude, the little chairs are a

virtual catalog of New Mexican tinsmithing designs and techniques. Both are constructed entirely of tin salvaged from cans and were originally decorated with oil-based paints.

The more complex chair on the left in Figure 4.38 was made from a can imported from Mexico, and portions of the label remain on the inside of the front skirt. Constructed with handmade tubing legs, back posts, and spindles, the chair features a one-piece embossed and scored panel that is bent to form the front and side skirts and the seat. The pieced back splat and top rail are made from an additional panel. Designs in both panels are related to those seen in the work of various New Mexican tinsmiths. For example, the crest of the top rail is similar to some lunettes made by the Fan Lunette Tinsmith, while the scalloped side skirts are scored and embossed much like the lunettes that are typical of the Santa Fe Federal Workshop. A scored eight-pointed star, similar to a detail found on work from the Rio Arriba Workshop, adds strength and a decorative design to the flat seat panel (see Chapter Five). The scrolled arms, along with the pierced and scalloped bottom stretchers, have been formed with a tinsmiths' combination rolling machine, a tool that was occasionally used by Hispanic tinsmiths in the early twentieth century. With the exception of the unpainted arms, the chair was originally painted entirely green, but most of the decorative parts have now been covered with silver paint.

The design of the more simple chair on the right in Figure 4.38 was derived from a Colonial Revival style commercial prototype. Narrow tin tubes form the legs, back posts, stretchers, scrolled arms, and back spindles. The shaped back rail is made from a flat sheet of tin decorated with single-dot embossing and circular stamping. The seat has a scalloped and pierced front skirt and a scored six-point star motif, while cone-shaped finials cap the back posts. This chair was originally painted green with red details but has now been overpainted in blue. Probably made by a tinsmith for his children or grandchildren, these beautifully scaled miniature chairs are very likely the only surviving examples of their type. Although diminutive, they show a New Mexican tinsmith's inventive skill and ingenuity.

Extremely few handmade pie safes with punched tin panels—generally common in the American South and Midwest during the nineteenth century—have been located in New Mexico. It is likely that midwestern commercial pie safes which came by railroad to New Mexico after 1880 were rarely adapted by the native craftsmen, but simply replaced local trasteros.[14] Two stamped tin sheets have been located that were salvaged from one of these factory-made midwestern pie safes. The centers of the sheets have been pierced to transform them into frames for religious prints.

In this study we have been concerned with the folk-art forms of the Hispanic tinsmiths and have not discussed basic functional products such as buckets, gutters, or roofing. Certainly, a few other decorative forms remain undiscovered or have been lost to time and the changing needs of the twentieth century. Perhaps some of these objects will eventually surface to give us a more complete understanding of the tinsmith's art.

PLATE 1
*Painted tin frame.
The Isleta Tinsmith,
circa 1900, 31 ×
25 1/2. Spanish
Colonial Arts Society,
Inc. collection on
loan to the Museum
of New Mexico,
Museum of
International Folk
Art, Santa Fe. All
dimensions are in
inches.*

44

PLATE 2
Sconce, one of pair.
Rio Arriba Painted
Workshop, circa 1890,
18¹/₂ × 11³/₈ × 6.
Private collection,
Santa Fe.

PLATE 3
Cross on base.
Possibly Valencia Red
and Green Tinsmith,
circa 1880. Cross:
17¹/₂ × 12³/₄; Base:
2¹/₂ × 2¹/₂.
Collection of Mary
Veitch Alexander,
Mesilla.

46

PLATE 4
*Document box with
hand-painted paper
panels. Possibly H.V.
Gonzales, circa 1890,
9 × 12¹/₂ × 6.
Private collection,
Santa Fe.*

PLATE 5
*Nicho. Rio Arriba
Painted Workshop,
circa 1890, 26 × 15
× 4. Spanish
Colonial Arts Society
Collection on loan to
the Museum of New
Mexico, Museum of
International Folk
Art, Santa Fe.*

48

PLATE 6
*Nicho with embossed chromolithograph.
Rio Arriba Painted Workshop, circa 1895,
$11\frac{1}{2} \times 11 \times 4$.
Private collection,
Santa Fe.*

PLATE 7
*Frame with
chromolithograph,
"La Face du Jesus."
Mora Octagonal
Workshop, circa 1875,
16³/₈ × 14³/₈. Kit
Carson Foundation,
Taos.*

PLATE 8
*Nicho with
chromolithograph.
José María Apodaca,
circa 1900, 13 × 11
× 2¹/₄. Private
collection, Santa Fe.*

PLATE 9
*Nicho. José María
Apodaca, circa 1900,
20¹/₄ × 16 × 4.
Alexander Girard
Folk Art Collection in
the Museum of
International Folk
Art, a unit of the
Museum of New
Mexico, Santa Fe.*

52

PLATE 10
Polychrome nicho.
Rio Abajo Workshop,
circa 1885, 32¹/₂ × 24
× 11. Collected in
Valencia County,
New Mexico.
Collection of Shirley
and Ward Alan
Minge, Casa San
Ysidro.

PLATE 11
*Frame with mirror.
Valencia Red and
Green II Workshop,
circa 1880, 22
diameter. Private
collection, Santa Fe.*

54

PLATE 12
*Frame with
oleograph. The Isleta
Tinsmith, circa 1905,
19³/₄ × 18.
Collection of Joan
Caballero, Santa Fe.*

PLATE 13
Frame containing Benziger oleograph. The Isleta Tinsmith, circa 1905, 31¹/₄·× 22¹/₂. Private collection, Santa Fe.

56

PLATE 14
Nicho. Santa Fe
Federal Workshop,
circa 1865,
Harwood Foundation
Museum, Taos.

PLATE 15
*Frame with painted
retablo on tin.
Mexico, circa 1875,
$10^{1}/_{2} \times 8$. Collection
of New Mexico State
University Art
Gallery, Las Cruces.*

58

PLATE 16
Nicho (ofreta).
Puebla, Mexico, circa
1885, 31¹/₄ × 14³/₄ ×
9¹/₂. Collection of
Robin and Barbara
Cleaver, Santa Fe.

5
THE TINSMITHS

*I*n our survey of more than one thousand pieces of New Mexican tinwork, we have been astonished at the small number of clearly recognizable workshops represented by these pieces. This suggests that a limited group of craftsmen produced vast numbers of tin pieces in the last half of the nineteenth century. In spite of the small population and the extremely remote locations of many of the villages of northern New Mexico, virtually every community must have had numerous tin objects hanging in homes, family oratorios (chapels), Penitente moradas, and the village church. We speculate that as many as five to ten thousand pieces may have been made in the course of the century. It seems logical that tin articles found in more remote areas of New Mexico were made by traveling craftsmen much like the itinerant santeros of the first half of the nineteenth century. Small villages like Llano Quemado, or Tres Piedras could not have supported a full-time tinsmith, though more populated communities such as Santa Fe, Bernalillo, Los Lunas, and Ranchos de Taos could have maintained resident craftsmen with permanent workshops. Larger towns, because of their role as trade centers, had access to the materials needed to produce tinwork: cans, solder, wallpaper, and glass. Some remote villages may have had access to discarded materials from U.S. Army forts nearby.

The workshops have been named for their possible locations on the basis of collection history and a distinguishing feature of the work. Most workshops developed more than one style of tinwork. Such differences in styles from a single workshop may have been caused by one craftsman working over a long period of time; all craftsmen produced a range of designs and over time changed from one design concept to another. The differences could also have been the result of two or more family members working side by side with differing ideas and skills. Thus, a wide range of design motifs are included within a single workshop. These multiple styles have been linked by the use of identical stamps, forms, and decorative details.

In the United States censuses from 1850 to 1910 there are more than twenty Hispanic men whose occupation is listed variously as tinsmith or tinner.[1] They lived in five counties of New Mexico, and were of both Spanish New Mexican and Mexican birth. Though their names are known, with two exceptions, there is scant evidence to link them to a specific body of work. The great majority of tinwork in New Mexico was produced by as few as fourteen workshops representing about twenty-five craftsmen with varying degrees of talent and skill. It is curious that there are nearly the same number of tinsmiths in the census returns as there are identifiable styles. Unfortunately, only one signed nineteenth-century piece has been found to date, though Revival period work is occasionally signed. Other tinworkers may have been listed as silversmiths or blacksmiths in the early censuses. Surely these craftsmen also had the technical skills to produce tinwork with their knowledge of tool making, soldering, and stamping. In the 1850 census, Santa Fe boasted thirteen silversmiths and as many blacksmiths among its craftsmen.

Some craftsmen may have been listed as farmers or laborers but during the long winters found the time to produce the tinwork needed in the community. Writing about her childhood in Arroyo Hondo, New Mexico, at the turn of the century, Cleofas Jaramillo mentions a tailor from Abiquiú, José María Baca, who made tin sconces and frames for her grandmother.[2]

We have been frustrated by a lack of connection between the names in the censuses and the workshops. Missing, too, are tinwork pieces with a collection history in areas where known tinsmiths worked, such as San Miguel and Mora counties in northern New Mexico. Who were these craftsmen who produced such a quantity of unique American folk art in such a short span of time? We hope that further research will reveal more names of makers of specific work.

Santa Fe Federal Workshop

The Santa Fe Federal Workshop pieces represent the earliest style found in New Mexican tinwork. Composed primarily of frames, this body of work exhibits meticulous craftsmanship and inventive design. Typical features are extensive and carefully executed stamping, diagonally scored engaged columns, and square corner bosses. These strong architectural motifs are derived from French Empire and American Federal period frames. The name *Santa Fe Federal* has been chosen because of the direct stylistic influence of the Federal prototypes and the long history of collection in Santa Fe County.

Dating Santa Fe Federal work is difficult. Few examples have survived which retain their original devotional prints or Holy Cards. Furthermore, the tinsmiths used only the unstamped portions of the discarded tin cans, leaving no clue to the origins or dates of the cans. The work may predate the common use of labeling on cans. The existence of a few Currier and Ives prints dated 1848–72 and European lithographs and Holy Cards from 1850 to 1870 suggest a date of design and manufacture between 1840 and 1870.

Santa Fe Federal frames are rectangular shapes formed of narrow side and end panels, often rolled and scored to represent engaged columns or ornamented with multiple designs of complex stamping and embossing. The panels are attached to square corner bosses commonly decorated with a rosette or a star-burst design. The frames are often surmounted by a crest in the form of a combined triangular pediment and a half-round lunette. Typically, the lunette is scalloped and scored to give the appearance of an open fan or a scallop shell, while the lower portion of the lunette extends the lines of a neoclassic pediment. With few exceptions, the frames are additionally ornamented with side appendages of both pendant and upright leaves.

Many examples of this work are simple and spare in decoration and design, following Federal prototypes. There are also numerous examples in which the basic frame has been elaborately ornamented with complex and fanciful shapes which create an elegant item of amazing beauty. In a few examples the crest is composed of a fanciful combination of stamped and embossed shapes, constructed as an elaborate crown. This type of construction resulted in pieces that were both delicate and fragile, and apparently very few examples have survived intact. The baroque character of these pieces is most closely allied to Mexican tinwork, whose prototypes were gold and silver baroque religious articles made during the Spanish Colonial period of the seventeenth and eighteenth centuries. The Santa Fe Federal tinsmiths were very likely familiar with tinwork being produced in Mexico in the second quarter of the nineteenth century as well as with the forms of both Empire and Federal wooden examples.

Although all the Santa Fe Federal pieces share strong stylistic and formal similarities, it is possible that the work may have been produced by more than one tinsmith. Some pieces are more sophisticated and elaborate than others, as though two tinsmiths were working in proximity or one was more creative and experienced than the other. Figures 5.1, 5.2, and 5.3 illustrate excellent examples of the Santa Fe Federal Workshop in its most complex style. Figure 5.1, intact except for its original print, represents all the typical characteristics of the Santa Fe Federal Workshop. The crest is a combined triangular pediment and fan lunette surmounted with fanciful accessory pieces. Ornamenting the sides of the frames are forms (pendant and upright) suggestive of oak leaves, but more closely tied to the rococo swags that decorated European gilt frames of the late eighteenth century.

CHARACTERISTICS OF
SANTA FE FEDERAL
WORKSHOP

5.1

5.2

FIGURE 5.1
*Frame. Santa Fe
Federal Workshop,
circa 1860, 28¹/₄ ×
20. Taylor Museum,
Colorado Springs,
Colorado.*

FIGURE 5.2
*Frame with oil on
canvas of St. John of
Nepomuk. Santa Fe
Federal Workshop,
frame circa 1860,
painting circa 1835,
27¹/₂ × 21¹/₂.
Collection of Shirley
and Ward Alan
Minge, Casa San
Ysidro.*

It is interesting to note that similar forms appear occasionally in the later work from the Isleta Workshop.

Figure 5.2 is an example of complex Santa Fe Federal style tinwork. The side panels of the frame are scored to resemble engaged columns tied together by four square corner bosses, each carefully embossed with rosettes. The simple crowning lunette lacks the fanciful additions of the previous example, though there is evidence of missing finials. Scrupulous craftsmanship is evident in the

execution of the six-pointed star and the scalloping of the lunette. The sides are ornamented with matching upright and pendant leaf shapes, and the scoring of the columns is regular and measured. The frame contains a Mexican painting of St. John of Nepomuk on linen, circa 1820–35. If the painting is original to the frame, this would be one of the earliest surviving examples of New Mexican tinwork.

Figure 5.3, although damaged and with its original print missing, clearly shows the creativity and

5.3

5.4

craftsmanship of the tinsmith. The end and side panels are decorated with carefully executed stamping and embossing, and the designs in the square corner bosses are a graduated series of delicate concentric rosettes. The crest is highly mannered with its elongated shape and unusual shell form. The one remaining fan-shaped appendage suggests that this piece was originally quite ornate.

Figures 5.4 and 5.5 are two of the simplest examples of the Santa Fe Federal Workshop. Both frames are fabricated from narrow strips of tin rolled

and scored to represent the typical columns connected by square corner bosses. The simplicity of the frames allows the devotional prints to assume greater importance than those dominated by more complex frames. Figure 5.4 is surely by the same hand as the three previous examples. Although stylistically similar, the frame in Figure 5.5 is more carelessly executed in basic shape and in decoration of the corner bosses. A peculiar energetic squiggle painting on the glass surrounds the French lithograph of "Je suis le servante du Seigneur." A pair of

FIGURE 5.3
Frame, incomplete. Santa Fe Federal Workshop, circa 1860, 22½ × 19. Private collection, Santa Fe.

FIGURE 5.4
Frame with lithograph of Our Lady of Solitude. Santa Fe Federal Workshop, circa 1855, 13 × 11. Collection of Ford Ruthling, Santa Fe.

5.5

5.6

FIGURE 5.5
*Frame with French
lithograph. Santa Fe
Federal Workshop,
circa 1865, 9¹/₄ ×
7³/₄. Private
collection, Santa Fe.*

FIGURE 5.6
*Frame with
lithograph of The
Sacred Heart of Mary.
Santa Fe Federal
Workshop, circa 1860,
21¹/₂ × 15¹/₂.
Collection of Shirley
and Ward Alan
Minge, Casa San
Ysidro.*

nichos by the same hand are in the collection of the Museum of International Folk Art. Constructed with glass painted with the same vigorous squiggles, they are rare examples of nichos from the Santa Fe Federal Workshop. The interiors have the customary engaged columns flanking small Benziger prints, circa 1865.

The delightful small frame in Figure 5.6, although delicate, possesses the same monumental proportions of the larger frames. The side panels in this example have been embossed with a group of small dots to create rosettes. The side leaf shapes are exceptionally charming, with their graceful recurved form repeated in lyre-shaped appendages that visually support the bottom of the frame. The typical lunette combines a triangular pediment with a scallop-shell motif radiating from a central rosette. The print is a French lithograph of "The Sacred Heart of Mary" made by Turgis, 1853–56.

The frame and image seen in Figure 5.7 possesses both aesthetic and historic interest. The frame contains a unique colored drawing of San Cayetano.

Now in the collection of the Millicent Rogers Museum in Taos, the frame formerly belonged to the San Ildefonso Pueblo potter, María Martínez. Similar in design to the previous frames, this one has extensively embossed and stamped side and end panels joined with fan-shaped corners. These corner pieces have been cut and folded at a 90-degree angle to create a three-dimensional form. Attached to the side panels are upright and pendant leaf shapes similar to those seen in Figure 5.1. The leaf shapes occupy the entire length of the side panels, which gives the frame an unusual sense of weight. The pendant leaf shapes are ornamented like those in Figure 5.2, but the lower leaf shapes are similar to those in Figure 5.3. The leaf on the lower left is a twentieth-century replacement made from terneplate. The crest is an unusual shape that retains some of the same ornamental vocabulary seen in previous examples. The embossed rosettes and curvilinear designs are produced with a series of small, closely placed, single-dot punches. These (along with the scored details) create leaf forms that are virtually repoussé.

Figure 5.8 is allied stylistically to the previously mentioned Rogers frame. Although less elaborate, the side panels are extensively decorated and painstakingly executed. While corner pieces and attached pendant leaf forms are typical of the workshop, the crest has not been seen in any other examples. The zigzag design repeats the ornamentation of the framing panels, and the scalloped edge of the lunette was meticulously measured and cut with a sharp chisel. The attached candle holders and bobeches, which appear to be original, create a frame-style sconce. The frame contains a Mexican

5.7

woodcut of the Santo Niño de Atocha and accompanying *alabanzas* (religious praises).

Two other frames with characteristics of the Santa Fe Federal Workshop are shown in Figures 5.9 and 4.1. Figure 5.10, another example, is simple in design yet elegant in form. Now fitted with a mirror, this frame is stylistically similar to the one seen in Figure 5.2. It is constructed of convex panels scored diagonally to represent columns. The joints are mitered at the corners, with little care taken to make the scored lines match. The scalloped lunette

FIGURE 5.7
Frame with hand-drawn and colored image of San Cayetano. Santa Fe Federal Workshop, circa 1865, 16 × 15³/₄, lower left side leaf form replaced. Millicent Rogers Museum, Taos.

5.8

5.9

is stamped and embossed much like the two side fins attached to the upper third of each side panel. The stamping on these frames is similar to that on some Rio Arriba Painted tinwork. It is possible that the two workshops may be related through a common workshop setting or the influence of one craftsman upon another.

The frame in Figure 5.9 is unusual, as it does not possess the airy elegance seen in other examples, and the overall stamping and embossing seem excessive and uncontrolled. This frame has the lap-

joint construction typical of the early Rio Arriba Workshop, but the surface ornamentation is more closely allied to Santa Fe Federal examples. Ornamental Gothic tracery frames the image in the print of Christ bearing the cross.

Figure 4.1 is an excellent example of the Federal style adapted by a less skilled craftsman. The frame is heavy and ponderous in comparison with the lighter and more elegant frame in Figure 5.1. E. Boyd's *Popular Arts of Spanish New Mexico* illustrates a related piece.[3] The form and side append-

5.10

5.11

FIGURE 5.10
Frame with replaced mirror. Santa Fe Federal Workshop, circa 1875, 20 × 18. Collections of the Museum of International Folk Art, a unit of the Museum of New Mexico, Santa Fe.

FIGURE 5.11
Sconce, one of pair. Santa Fe Federal Workshop, circa 1860, 17³/₄ × 17, bobeche has been replaced. Collections of the Museum of International Folk Art, a unit of the Museum of New Mexico, Santa Fe.

ages in Figure 4.1 appear as flying wings, while in the frame referred to by Boyd the side appendages are birds that resemble swans.

One of the most exceptional pieces made by the Santa Fe Federal Workshop is the sconce illustrated in Figure 5.11. One of a pair, it is a composite of all the details of the previous frames. The main body of the sconce contains a six-pointed star with ornamentation surrounding it identical to that in Figure 5.6. The wings are similar in form and execution to the pendant leaf forms in Figure 5.1. The fluted crown above the double eagle heads is a variation of the crest in this same piece. A Federal-style mirror brought over the Santa Fe Trail could have been the design inspiration for the lovely sconce, or perhaps it was a Mexican frame (ornamented with the Hapsburg double eagle) that was brought up the Chihuahua Trail. The double-headed eagle motif was introduced into Mexico from Europe and remained popular into the early twentieth century.

The work of the Santa Fe Federal Tinsmith, then,

represents the most elegant, sophisticated, and well-designed tinwork produced in mid-nineteenth-century New Mexico. It is fortunate that so many of these early examples have survived.

Rio Arriba Workshop

We have designated a prolific workshop of the last third of the nineteenth century the Rio Arriba Workshop. (In this context Rio Arriba simply refers to northern New Mexico.) The tinsmiths in this workshop, who probably lived in or near the Española valley, produced a large number of frames, nichos, and miscellaneous items, many of exceptional quality. Although the work has been scattered throughout New Mexico, Colorado, and California, a large group has remained in northern New Mexico, primarily in the Española valley and in Santa Fe. Two examples are known to have been collected in the village of Chimayó, and others have been collected at San Juan Pueblo adjacent to Española in Rio Arriba County.

Referring to examples from this workshop that contain wallpaper, notes of the late E. Boyd date the wallpaper 1870–90.[4] Many of the framed prints date from this same time. Can labels from the Colorado Packing and Provision Company (after 1893) suggest a period of activity for the workshop from about 1870 to 1895.[5]

The Rio Arriba Workshop produced simple frames made from single pieces of tin salvaged from flattened tin cans and cut into rectangular or round shapes. A pierced rectangular or square opening accommodated the small devotional print or Holy Card. In addition to commonly used devotional prints, the tinsmiths occasionally used greeting cards or religious catalog advertisements for commercially produced plaster figures of Catholic saints. At times, the printed images were backed with small pieces of wallpaper or decorative wrapping paper which served as a border. The frame's rectangular opening was usually accented with lines produced by scoring. This scoring seems haphazard, with little or no attempt to make lines parallel or to conform to the opening. The simple rectangular or round frames often have scalloped borders or rounded corners. The surfaces are ornamented with widely spaced, random stamping of an open multipoint star produced by a single serrated, circular stamp. The empty center of the star is embossed with a single-dot punch.

The Rio Arriba Workshop also produced complex rectangular frames cut from four pieces of tin and assembled with mitered or butt joints much like eastern Federal-style frames. There are examples of simple frames in the Santa Fe Federal style which bear stamping similar to that found on later pieces. This suggests that a Rio Arriba tinsmith may have begun his craft earlier than our presumed dates, at which time he produced frames with side leaf forms and simple lunettes. Figure 5.12 is similar in design to frames which appear in a photograph dated 1872 that shows the interior of the Church of the Holy Cross in Santa Cruz.[6] The frames appear to be by the Santa Fe Federal Tinsmith. It is quite possible that a Rio Arriba tinsmith would have known of the Santa Fe Federal workshop and would have been influenced by the designs. It is also very likely that there was more than one Rio

CHARACTERISTICS OF
RIO ARRIBA WORKSHOP

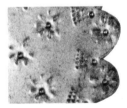

5.12

5.13

Arriba tinsmith, which accounts for the variety of craftsmanship and design.

Similar in construction to frames, nichos utilize individual pieces of tin to form decorative side panels. In some nichos, tin side panels were replaced with strips of wallpaper protected by narrow glass panes held together with narrow tubes of tin. The use of wallpaper panels in place of tin is especially common in larger nichos made to contain bultos.

Identifying characteristics of the Rio Arriba Workshop include the extensive use of the circular serrated stamp and reverse single-dot punch. This stamp is used singly, in rows, or in groups of five units. A serrated U stamp is frequently used, either reversed to produce a meander pattern or employed in groups of four, producing a quatrefoil design. In all instances, the U stamp is combined with the embossed single-dot punch. Other designs and patterns frequently used include small multipointed stars. The formation of the star is made by an individual strike of a serrated bar stamp, the stars often connected to the serrated circular pattern by

FIGURE 5.12
Frame with devotional page. Santa Fe Federal Workshop, circa 1870, 17 1/4 × 16. Collection of Ford Ruthling, Santa Fe.

FIGURE 5.13
Nicho. Rio Arriba Workshop, circa 1880, 24 × 17 1/2 × 5 1/2. Collections of the Museum of International Folk Art, a unit of the Museum of New Mexico, Santa Fe.

popular motifs of the Gothic Revival style of the late nineteenth century. In addition to the repoussé stars and birds (found exclusively on lunettes), smaller birds often are cut from tin and soldered around the perimeter of the lunette and on the top of the finials. Such bird designs, not easily identified as any particular bird, are unique to this workshop. They are commonly outlined with single-dot punching and ornamented with stamping and reverse single-dot punching scattered throughout the body of the bird (see Figure 5.13).

An applied ornamental device found most commonly on nichos—but infrequently on frames—is the use of scalloped leaf forms suggesting oak leaves. Other side appendages that also appear are semicircles with the interior cut to form an arch. These are also extensively ornamented with stamping and single-dot embossing (see Figures 5.18 and 5.19).

Both Figures 5.14 and 5.15 illustrate good examples of simple frames containing devotional prints. The piece in Figure 5.14, well constructed and stylistically similar to earlier rectangular Federal frames, is decorated with stamping typical of the Rio Arriba Workshop. The edges are impressed at regular intervals around the border with a small serrated bar stamp. The interior of each side and end panel is ornamented with a serrated crescent stamp that creates a scalloped pattern. The center of each panel contains the serrated circular stamp connected to a larger serrated bar, producing a Greek fret pattern. The entire frame is slightly convex, offering a three-dimensional quality not usually found in this workshop. The well-preserved Benziger two-color lithograph dates from the late nineteenth century. Both the late date of the print and

5.14

a large-toothed straight bar punch. Such a combination creates a Greek fret image. This pattern appears most frequently on the more ambitious, carefully executed frames and on side and end panels of large nichos.

Elaborate nichos often have a large five- or six-pointed star on the lunette, produced by single-line scoring and repoussé. The star is usually flanked by a pair of fat, whimsical birds (see Figure 5.18). Other identifying characteristics are floral rosettes and scalloped finials, which may be a response to

FIGURE 5.14
Frame with Benziger lithograph titled "S. Vincentius A Poulo." Rio Arriba Workshop, circa 1880, 14¹/₂ × 12¹/₂. Collection of Ortíz y Pino family, Galisteo.

5.15

5.16

FIGURE 5.15
*Frame with Benziger
Holy Card. Rio
Arriba Workshop,
circa 1875, 8¹/₂
diameter. Spanish
Colonial Art Society,
Inc. collection on
loan to the Museum
of International Folk
Art.*

FIGURE 5.16
*Frame with Benziger
lithograph of St. Ann
and the Virgin Mary,
mounted on
wallpaper. Rio Arriba
Workshop, circa 1880,
18 × 12¹/₂.
Collection Ortíz y
Pino family, Galisteo.*

the careful craftsmanship suggest that this speci-
men was crafted at the height of the craftsman's
productive period.

Figure 5.15 shows one of many round frames in
existence. The edges are scalloped, each scallop ornamented with a single serrated circular stamp; the
interior surface is decorated with like stamping. An
additional die used is an unexpected triangular shape
that is infrequently seen in work from this workshop. The centered rectangular opening contains a
Benziger print of Jesus and St. John as children. The

opening is outlined unexpectedly with crude single-
line scoring. The stamping occurs at regular and
calculated intervals, but apparently little care was
taken to ensure evenly measured scored lines.

Figure 5.16 shows a good example of greater care
being taken in the execution and design of a frame.
This scalloped rectangular shape is decorated with
the familiar serrated circular stamp. The scoring
that accents and outlines the opening has been carefully executed. Two side finials and a circular crown
make the frame seem more complex. The print, a

5.17

5.18

FIGURE 5.17
Nicho containing painting on canvas of Nuestra Señora de Guadalupe. Rio Arriba Workshop, circa 1880, 22³/₄ × 20¹/₂ × 5¹/₄. Millicent Rogers Museum, Taos.

FIGURE 5.18
Nicho. Rio Arriba Workshop, circa 1880, 24 × 19 × 4. Formerly in Randall Davey Collection. Collection of Mr. and Mrs. Courtlandt Barnes, Santa Fe.

delicate Benziger lithograph of St. Ann and the Virgin Mary, is placed against a fragment of an Arts and Crafts period blue wallpaper.

The Rio Arriba Workshop tinsmiths seem to have been truly inspired and creative when making nichos. Figure 5.17 illustrates an excellent example of one of numerous large nichos. The form displays a good sense of proportion and scale. Side and end panels are constructed of glass and a beautifully patterned green wallpaper. The sides are ornamented with scalloped wings decorated with the

ubiquitous serrated circular stamp, and the simple lunette is decorated with floral rosettes and stars. Four quarter circles form the corners of the nicho that tie the side and end panels together. These are ornamented with the quatrefoil design.

Other large nichos are illustrated in Figures 5.13, 5.18, and 5.19. The nicho in Figure 5.13 is decorated with a type of stamping often seen on frames. In this instance, the U stamp has been reversed to define an S pattern, appearing in a repeated sequence along the side and end panels of the body.

The large, central circular crown also contains U stamping, which is combined with the large serrated bar stamp to form teardrop shapes arranged in a fanciful pattern. The light and delicate quality is further enhanced by the inverted oak-leaf finials and side pendant leaf forms. As pure ornament, small birds surround the entire nicho. The image of these birds seems unrelated to the image of the Holy Spirit. Indeed, the two birds at the apex of the crown seem to be engaged in serious conversation.

The exquisite nicho illustrated in Figure 5.18 is an example of the Rio Arriba Workshop at its best. The nicho is fabricated in the common trapezoidal form, with the addition of appendages and a lunette providing visual interest. The square corner bosses each contain carefully stamped rosettes; the somewhat restrained nature of these corner bosses is a nice contrast to the elaborate composition of the nicho. All the edges of the side panels, lunette, and flaring semicircular appendages have been scalloped repeatedly with a small crescent-shaped cutter, creating an overall shimmering visual quality. The lunette crowning this superb work contains the familiar charming birds and the six-pointed star that is frequently present in the work of the Rio Arriba Workshop—the birds and star created with single-dot stamping, scoring, and repoussé. The carefully measured cutting, stamping, and embossing of each element of this nicho attest to the skill of the tinsmith and show his eye for design, proportion, and detail. While tinwork from this workshop is typically devoid of color, we have occasionally found painted examples. Most of the painted items appear to have been painted subsequent to their fabrication. An exception is this well-preserved ni-

5.19

cho, which retains most of its soft yellow and green color. The paint has been carefully applied in a manner harmonious with the overall design of the nicho.[7]

A nicho probably from the Santuario de Potrero near Chimayó, New Mexico, is illustrated in Figure 5.19. This extremely large nicho and an accompanying chandelier were collected near Chimayó in the 1950s. If the nicho was made for the santuario, its large size plus the presence of two tubes attached to the underbody suggest that it was made to be carried on wooden poles in religious processionals.

FIGURE 5.19
Nicho. Rio Arriba Workshop, circa 1885, 40 1/2 × 34 × 10. Private collection, Santa Fe.

5.20

FIGURE 5.20
*Chandelier. Rio
Arriba Workshop,
circa 1875, 15¹/₂ ×
23¹/₂ diameter.
Private collection,
Santa Fe.*

Now missing its bulto, the nicho must have been an impressive sight as it was carried to its place of honor on the altar.

The Chimayó nicho is constructed of large sheets of glass which form the front and sides, while the back is tin sheet from a single large can. The primary decorative forms consist of extensively stamped and embossed central lunette, two smaller lunettes, and six semicircular arcs attached to the side panels. The design of the main lunette is almost identical to that of the previous example, with the exception of the scalloped edges. Originally there were smaller birds attached to the lunette, but these have been lost. The tin border that frames the large glass panels is decorated with repetitive quatrefoils.

The chandelier (Figure 5.20), also collected with the nicho, is one of few surviving from the nineteenth century, and although it has a rather clumsy appearance, it is actually sophisticated in design. It is constructed with a double row of circular tin bands, one supporting candle holders. The bands as well as the connecting strips have been stamped with the familiar serrated circular stamp and a large serrated bar stamp, forming a teardrop motif. The whole piece has been painted the same unexpected blue as the nicho. One can imagine this chandelier hanging in the transept, providing an elegant source of illumination.

The Rio Arriba Workshop pieces may have been made by one man during a period of perhaps twenty years, or it is possible that they were the production of two or more craftsmen working together (or in close proximity), sharing ideas and a similar design vocabulary. In either circumstance, the work shows a definite progression from rudimentary execution in form and design to a sophisticated degree of craftsmanship and complex design. The best tinwork from this workshop ranks among the finest produced in northern New Mexico in the last quarter of the nineteenth century.

The Taos Serrate Tinsmith

Taos Serrate represents a small group of pieces that are energetic in execution, sometimes large in scale. The style was named for Taos County, where the pieces have most commonly been collected, and refers to the prominent edge detail. The work is easily distinguished from other New Mexican tinwork of the late nineteenth century. The most prominent characteristic is a consistent use of serrated borders on side and end panels and lunettes of both frames and nichos. Two further characteristics are readily recognized. First, the back panels of nichos and sconces are often ornamented with a band of multiple chevrons produced by repeated stamping with a toothed bar stamp. The stamping and embossing are deep, and at times rather clumsy. Occasionally, the entire back panel is so filled with bands of chevrons and other motifs that the surface resembles repoussé. Second, simple rosettes often accompany the chevron motif. They are created by stamping with a toothed crescent tool around the perimeter and embossing with a toothed bar stamp to produce a star-burst pattern within the scalloped rosette (see Figure 5.23). Although extensively ornamented and heavily stamped, the ornate Taos Serrate frames characteristically utilize only three stamps: the single-dot punch, the serrate-toothed crescent stamp, and the serrate-toothed bar stamp. (Occasionally, a very small six-point star stamp is also used.)

Dating this work is difficult, because the few surviving can labels do not show enough information to verify them with accuracy. With few exceptions the work does not contain wallpaper or other datable materials. Surviving original prints include French lithographs dating from the 1870s and chromolithographs from the 1890s. Based on these and other collection data from the examples which will be discussed, the period of production was 1870 to 1905.

The nicho in Figure 5.21 may be considered the best surviving example of the work from the Taos

5.21

Serrate Workshop. It incorporates all the typical characteristics found separately in other works of this group. The basic form of the nicho is a rectangular frame with attached finials and serrated lunette. The perimeter has been cut to produce a large serrated border, while the projecting body of the nicho is trapezoidal. The upper and lower serrated border panels have been wrapped around the floor and ceiling of the nicho, giving it the appearance of an architectural structure. The panels are extensively stamped and embossed, primarily with the

CHARACTERISTICS OF
THE TAOS SERRATE
TINSMITH

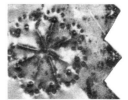

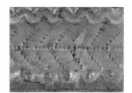

FIGURE 5.21
Nicho. Taos Serrate Tinsmith, circa 1885, 27 1/8 × 19 3/4 × 5. Collection of Sheila and Julian Garcia, Albuquerque.

5.22

5.23

FIGURE 5.22
Frame with double portrait. Taos Serrate Tinsmith, circa 1901–5, 12¹/₂ × 14¹/₂. Collection of Mr. and Mrs. Murdoch Finlayson, Santa Fe.

FIGURE 5.23
Sconce. Taos Serrate Tinsmith, circa 1885, 10 inches high. Collection of Tony Garcia.

repeated chevrons. In the lunette, the band of chevrons follows the outer curve to encompass three rosettes nicely composed in the center.

The finials, although not serrated, have a definite angular quality in keeping with the designs from this workshop. Each finial is also stamped with a rosette similar to those in the lunette. An unusual detail for this workshop is the addition of two birds perched atop the finials, each bird appearing to look inside the nicho. The back panel of the nicho displays the same typical stamped and

embossed ornamentation that defines the front and sides of the piece. Although it retains its symmetry, the decorative stamping of the back is allowed expressive freedom; the chevron band meanders within its confines and the rosettes appear to float, in contrast to the rigid design of the outer border.

Figure 5.22 illustrates a simple frame from the Taos Serrate Workshop. Side and end panels are not serrated, but contain characteristic rosettes. The sense of substance and weight is characteristic of all the pieces from this workshop. The two cabinet

5.24

5.25

cards represent a rare example of the use of photographs instead of religious images. They are mounted on a preprinted card from Mary Richardson of 878 Second Avenue, Durango, Colorado. Mrs. Richardson's studio was located at this address only in 1900 and 1901, after which she moved her business to Lake City, Colorado.[8]

Figure 5.23 shows a typical L-shaped nineteenth-century utilitarian sconce. Elaborate embossing and a serrated top edge make this piece easily recognizable. Decorative details are nearly identical to those in the side panels of the nicho in Figure 5.23. The Taos Serrate tinsmith generally was content to repeat the same decorative motifs and adapt them to each particular form.

The large sconce in Figure 5.24 is one of a set of seven collected from Casa Salazar, near Cabezón, New Mexico. Although it does not have a serrated border, it does exhibit the characteristic heavy-handed stamping and embossing seen in the nicho in Figure 5.21. The rosettes and radiating lines of chevrons float within the scalloped outline as though

FIGURE 5.24
Sconce, one from set of seven. Taos Serrate Tinsmith, circa 1880, 19³/₄ × 16¹/₄ × 4³/₄. Collection of Shirley and Ward Alan Minge, Casa San Ysidro.

FIGURE 5.25
Sconce, one of set of seven. Taos Serrate Tinsmith, circa 1880, 15 × 11 × 7³/₄. Private collection, Santa Fe.

5.26

FIGURE 5.26
*Frame with Holy
Card and wallpaper.
Taos Serrate
Tinsmith, circa 1885,
10¹/₂ diameter.
Collections of the
Museum of
International Folk
Art, a unit of the
Museum of New
Mexico, Santa Fe.*

enveloped in an airy, undefined space. Cut from one large piece of a salvaged can, the shape is basically rectangular. The addition of semicircular pieces around the perimeter results in an oval shaped sconce, however. These sconces are some of the largest surviving examples from New Mexico.

A third sconce, seen in Figure 5.25, although much more elaborate and imaginative in design, incorporates many decorative elements and design motifs similar to those seen in Figure 5.24. This sconce is one of seven collected about 1920 by Miss

Amelia Hollenback of Santa Fe. According to Miss Hollenback's records, the sconces were collected from the church of San Ysidro in Corrales, New Mexico.[9] It is more probable, however, that they came from the church in San Ysidro, near Cabezón, Sandoval County, New Mexico. By employing a large expanse of extensively ornamented tin surrounded with additional pieces of stamped and embossed tin, a series of reflective surfaces were created. This caused light from a single candle to be refracted and diffused, producing additional illumination.

The two pieces illustrated in Figures 5.24 and 5.25 are representative examples from larger sets collected in small mission churches. They bear a close resemblance to other work by the Taos Serrate Tinsmith, particularly in their deep chevron and rosette style of stamping. These sconces do not have the typical serrated edges, however. The sets were collected in Cabezón (now a ghost town) and San Ysidro, in an extremely remote area of northwestern New Mexico. These villages were important stopping points on the stagecoach line going west in the 1870s from Santa Fe to Fort Wingate.[10] We speculate that these pieces may have been shipped to the area or fabricated by an itinerant tinsmith on the site. Their scale and form are unlike those of the other work by this tinsmith and may represent earlier work made before the mature style developed, pieces made by a related craftsman, or pieces purposefully made large to fit the churches.

Figure 5.26 shows a lovely example of a circular frame. Cut from a single piece of tin, the design is reminiscent of small round frames from the Rio Arriba Workshop. The frame has the typical serrated border and familiar embossed rosettes. The

5.27

5.28

FIGURE 5.27
Frame with Turgis chromolithograph of Saint Dominic. Taos Serrate Tinsmith, circa 1885, 10$^1/_2$ × 10$^1/_4$. Spanish Colonial Art Society, Inc. collection on loan to the Museum of International Folk Art.

FIGURE 5.28
Nicho-frame with tintype. Taos Serrate Tinsmith, circa 1885, 13$^1/_4$ × 11 × 2$^1/_2$. Collection of Will and Deborah Knappen, Santa Fe.

pierced rectangular opening frames a small colored devotional print of the Holy Family and is set against a piece of handsome Victorian wallpaper.

Figure 5.27 is the most unusual style of frame from this workshop. It bears the familiar serrated lunette complete with embossed and stamped rosettes. The simplified frame is constructed of tin tubing designed to hold the Turgis devotional image of St. Dominic and the glass securely in place. The elongated side scrolls attached to the tube frame are an odd shape but are not unique to this piece.

Similar examples have been found and are also embossed with the rosette.

One of the most beautiful and certainly one of the most charming examples from the Taos Serrate Tinsmith is the framed shallow nicho in Figure 5.28. This piece is unusual in that it is mostly glass with few tin details. The only recognizable characteristic which identifies it as the work of the Taos Serrate Tinsmith is the set of four fan-shaped corner pieces connecting the glass side panels. The corners are cut with large serrated edges and stamped with the familiar rosette. The side and end panels are held together by narrow tubular tin pieces, and the center is a projecting shallow nicho. The interior space is entirely too small and shallow to have housed even the smallest bulto. Rather, it was undoubtedly built expressly for this framed tintype of a bulto. The delicate pastel wallpaper under the glass panels is in the Victorian style. This is consistent with the few other examples of Taos Serrate tinwork that employ wallpaper as a part of the decoration. The photograph is surrounded by a printed decorative blue border from a Benziger lithograph. This most unusual and charming nicho is a testament to the tinsmith's creative and loving care.

Few of these immensely appealing Taos Serrate pieces have been located, although others must exist. It is baffling that such a distinctive body of work, collected in both Taos and Sandoval counties, should occur in such widely separated locations.

Rio Arriba Painted Workshop

A productive workshop we have titled Rio Arriba Painted produced work characterized by fanciful frames and elaborate nichos, sconces, and trinket boxes. A distinctive feature of this work is the extensive use of reverse-painted panes of glass. With few exceptions the images are floral motifs of roses or tulips. The most common colors are blue, yellow, yellow-green, green, red, and pink. Occasionally the tinsmiths employed a monochromatic color scheme, but it is more common to find pieces painted with red, yellow, green, and blue. The name was chosen based on the collection history centered

in Rio Arriba County and the consistent use of the distinctive painted glass.

Rio Arriba Painted nichos are large and elaborate, occasionally incorporating Gothic Revival–influenced architectural elements along with theatrical lighting techniques. While painted glass panels are the predominant feature of nichos (and to a lesser extent frames), there are frames made entirely of tin. These are decorated with a style of stamping unique to this workshop.

The most characteristic elements are two stamps; one produces a pair of impressions suggestive of deer tracks, while the other has four inverted V shapes that give the effect of a Greek cross. There is also a serrated crescent stamp distinct from those found in other workshops. Another image, usually embossed, is similar to the Greek cross design. Stamped into the tin surface, the die produces four inverted diamond points. The combination of a toothed crescent die and toothed bar stamp results in a large teardrop. With variations, this motif is found consistently in the work of the Rio Arriba Painted Workshop.

Collection information points to a northern New Mexican origin for this work. One possible source comes from Cleofas Jaramillo's stories of her childhood in Arroyo Hondo, written in the late 1930s. Mrs. Jaramillo describes a tinsmith working at her grandfather's home near the end of the nineteenth century:

At grandpa Vicente's house there was work even during the bleak winter months. In November, word was sent to Don José María Baca, the tailor from Abiquiú, to come to outfit the family. . . . The tailor had to make suits for the entire family, including the

servants. . . . This work finished, Grandma put the tailor to work, making fancy tin sconces adorned with little mirrors, surrounded by bits of colored glasses. He also made lovely tin frames for Grandma's collection of *santos*, donated to her by the bishop and priests who stopped at the house during the year.[11]

The 1880 and 1900 census returns for Rio Arriba County have been examined for a reference to José María Baca. There is neither a listing for anyone named Baca in Abiquiú during this period, nor for a tailor or a tinsmith. A José María Baca was listed in the 1880 census. He lived in El Rito, New Mexico, fourteen miles north of Abiquiú. Mr. Baca was sixty years old in 1880; his occupation was listed as laborer. Mrs. Jaramillo was recalling her childhood more than fifty years later, so it is possible that she was mistaken about Mr. Baca's home. There are some connections, however. Mrs. Jaramillo had relatives in El Rito, and her husband was born there. Several El Rito families still remember stories of grandfathers who produced tinwork during the winter months, probably about 1890.[12]

Mrs. Jaramillo's "bits of colored glasses" were in all probability a reference to painted glass panels which might have appeared to a small girl to be colored glass. No nineteenth-century examples of true colored glass have surfaced in a survey of more than a thousand pieces of New Mexican tinwork. The Rio Arriba Painted is the only style made in northern New Mexico that would be a possible source for this reference. There are at least two pieces of Rio Arriba Painted tinwork that point to an El Rito/Abiquiú origin. A small nicho with painted glass panels which was collected in El Rito in the 1930s from Manuel Martínez is now in the Wood-ard Collection at Adams State College in Alamosa, Colorado.[13] A cross with typical floral painted panels was found in an Abiquiu morada in the 1960s.[14] Although the Rio Arriba Painted tinwork presently cannot be attributed to José María Baca, the connections are intriguing.

Many examples of the Rio Arriba Painted Workshop no longer contain their original prints. There are examples of Currier and Ives, Benziger, and German lithographic prints and Holy Cards that date from the 1870s and 1880s. A conspicuous absence of large Benziger oleographs suggests that the tinsmith ceased work prior to 1890. If the maker was Mr. Baca of El Rito, he may have been active from 1850 to 1890, although the majority of this work dates from 1870 to 1890.

Figure 5.29 is an exceptional example of the tinsmith as painter. The frame is made up of glass panels combed and painted with roses. Thin tubes hold the glass panels together. The print of St. Rupert does not fit comfortably into the octagonal opening, however. An additional piece of glass has been fitted across the bottom, modifying the opening from an octagon to an arch. Even including the title information in three languages at the bottom, this print would have fit into the original octagonal space. Could the craftsman originally have designed the frame for a smaller print?

The imposing nicho in Plate 5 is related in concept to this frame. It is composed of eleven painted glass panels surrounding a three-dimensional construction of clear glass. The exterior frame is crowned by an elaborate lunette formed of elongated diamond-shaped painted glass panes. Conceived as an architectural space complete with pediment and

5.29

5.30

FIGURE 5.29
*Frame with German
chromolithograph of
St. Rupert. Rio Arriba
Painted Workshop,
circa 1890, 18³/₄ ×
14¹/₈. Spanish
Colonial Art Society,
Inc. collection on
loan to the Museum
of International Folk
Art.*

FIGURE 5.30
*Frame with
superstructure and
Currier and Ives
lithograph. Rio Arriba
Painted Workshop,
circa 1880, 17¹/₄ ×
14⁷/₈ × 3. Kit Carson
Foundation, Taos.*

pitched roof, this large nicho is a *tour de force* of construction. The clear glass roof functions as a clerestory, admitting light from above to illuminate the small framed print attached to the back panel. Why did the tinsmith invest such energy and creativity to embellish a diminutive print? The floral painting is enhanced by myriads of applied tin flowers. These are found attached to the tips of the sunburst lunette, floating around the framed print, and perched on top of the two curious cone finials flanking the pitched roof.

Similar in both its three-dimensional construction and its ornamentation of tin flowers is the frame in Figure 5.30. The unusual surface decoration is entirely embossed with three stamps: a large toothed crescent, a notched bar, and a notched circle. A unique aspect is the attached superstructure of tubular tin pieces that projects from the frame. Fourteen scalloped tin flowers are applied to the structure, and two fleur-de-lis finials are mounted at the top. A portion of a tin can embossed "nkinton & Ar" is visible on the back. This is part of a lard can from the Plankinton & Armours packing company of Kansas City dating from the 1870s. The Currier and Ives print of "The Holy Communion," showing a manner of dress fashionable from 1840 to 1865, was published in the early 1870s. This must have been a popular and widely distributed print, judging from the number of identical prints that have survived.

Similar in form is the simple frame with lunette in Figure 5.31. It is decorated with graceful stamping on the broad side and end panels. The row of notched crescent stamps is embossed with a stamp peculiar to the Rio Arriba Painted Workshop, made of four well-defined blunt triangles. Characteristic teardrop motifs appear in single rows in the framing panels and again in the scalloped lunette. The teardrops in the center of the lunette, adjacent to each other, produce a fan-shaped image. This decorative device is also found in Revival period tinwork, especially in the work of the late Robert Woodman (see Chapter 6). The two scalloped semicircles flanking the partially open fan in the lunette are found repeatedly in the stamping of a remarkable nicho in the collection of the Museum of International Folk Art.[15]

One of a pair, the sconce in Plate 2 was collected before 1929 by Santa Fe painter and collector Frank Applegate. This dramatic and delicate sconce exemplifies the fine craftsmanship and exquisite sense of design of the Rio Arriba Painted Workshop. The diamond shape in the center and the circular crest are carefully stamped and embossed with a rosette utilizing the deer track and Greek cross stamps unique to this work. Scored lines that radiate from the diamond in the circular crown are repeated in

5.31

the two side fans, and the L-shaped bottom bracket has a scored border and two embossed stars flanking the candle holder. The glass panels are back-painted with green and yellow. Combed flowers and leaves echo the ethereal quality of the linear arches, which are made from twisted tin strips. These strips, another detail commonly used by tinsmiths during the Revival period of the twentieth century, are rare in nineteenth-century tinwork.

The perfectly preserved nicho in Plate 6 is similar to these sconces. Here, the basic rectangular

FIGURE 5.31
Frame. Rio Arriba Painted Workshop, circa 1890, 21 1/4 × 14 1/2. Harwood Foundation Museum, Taos.

located.[16] Apparently, these nichos were avidly collected in the 1920s and 1930s. They were often electrified and set into plastered walls, to be used as sconces. Figure 6.13 illustrates a Revival period nicho by Pedro Quintana which is based on these examples.

The trinket box in Figure 5.32 features similar glass panels, painted and combed to resemble cabbage roses. In this example, painted panels are placed directly over the tin background instead of the more common white paper. The hutch shape is typical of nineteenth-century New Mexican boxes.

The deliberate choice of painting glass (rather than using wallpaper or cloth) and the use of theatrical lighting effects suggest that the Rio Arriba Painted craftsmen may have had contact with Mexican tinwork in the last half of the nineteenth century. Careful attention to detail and meticulous craftsmanship are clearly evident in the work of these craftsmen, who were some of the most imaginative and talented tinsmiths in nineteenth-century New Mexico.

5.32

Mora Octagonal Workshop

This workshop is named for its area of origin—Mora County—and for the shape of the most prominent examples of the work. It produced a variety of objects, including small crosses, massive nichos, and frames. Examples have been located in Taos, Santa Fe, and the Española valley as well as in San Miguel and Mora counties, all in northern New Mexico. A number of fine examples virtually identical to the frames in Figures 5.33 and 5.34 still

FIGURE 5.32
Trinket box. Rio Arriba Painted Workshop, circa 1890, 4⁵/₈ × 8¹/₄ × 5¹/₄. Private collection, Santa Fe.

form has been altered to a square. Side and end panels are made of reverse-painted glass in blue, green, yellow, and orange, all on a white ground. The active surface quality produced by combed painting is related visually to the scored and scalloped corner pieces. A rosette motif is stamped and embossed in the upper corners, while the lower scalloped cups are fitted with candle holders. The shallow interior contains an embossed lithograph of Our Lady of Carmel attached to the back of the nicho. This example is one of five similar ones

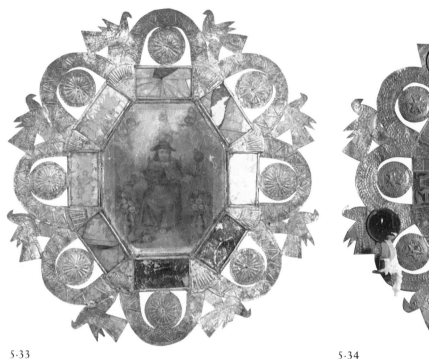

5.33

5.34

FIGURE 5.33
*Frame with
lithograph of Santo
Niño de Atocha. Mora
Octagonal Tinsmith,
circa 1870, 26 1/2 ×
25 1/2. Collection of
Ford Ruthling, Santa
Fe.*

FIGURE 5.34
*Frame-sconce with
lithograph and
wallpaper. Mora
Octagonal Tinsmith,
circa 1870, 27 1/2 ×
17 1/2. Former
collection of Dorothy
S. McKibbin.*

hang in homes and churches in Mora County.

An extensive collection now housed at El Rancho de Las Golondrinas in La Cienega, New Mexico, was formerly in the collection of José Ortíz y Pino of Galisteo, New Mexico. This accumulation includes the largest single group of Mora Octagonal pieces. The Ortíz y Pino family amassed the work primarily in San Miguel and Santa Fe counties in the late nineteenth century and the first half of the twentieth century.

The workshop is named specifically for the large octagonal frames in Figures 5.33 and 5.34. A fragment of a frame identical in design is said to have been collected from a Penitente morada in Rociada, New Mexico.[17] Similar octagonal frames still grace homes and churches in Mora County. Dating the work is difficult, for few can labels appear on the back of the frames and a broad range of printing techniques are found. Several frames contain devotional prints dating from the mid-1860s. There are also Benziger devotional prints and Holy Cards dating from the last third of the nineteenth century, as well as later chromolithographs and oleographs. Numerous frames and nichos contain wallpaper or border paper of both the Victorian and the Arts and Crafts periods. Based on dates for the prints and wallpaper, the period of production of the Mora Octagonal Workshop was 1860 to 1900. This forty-year period suggests that several closely allied tinsmiths, perhaps family members or neighbors, produced the work. One possible craftsman for some of these pieces is found in the census records. In the 1880 census, Juan Armijo was listed as a tinsmith living in La Cueva, Mora County, less than twelve miles from Rociada in San Miguel County.[18] The date and location in such a remote area suggest that Juan Armijo could be the maker of these pieces.

Mora Octagonal work has a number of distinguishing characteristics. The entire surface is frequently covered with extensive stamping and embossing so repetitive and closely placed that there is little negative space to act as a visual respite from the busy surface. The repetitive rows of stamping are either continuous C figures or a series of connected horseshoe stamps. The borders are scalloped with a rounded crescent cutter, the force of the punch as it cut the tin producing deep concave scallops that soften the edges visually. The large scalloped border also appears frequently on the lunettes and tin corner pieces of glass paneled nichos. Wallpaper panels are a significant detail used by this workshop, and serve as the basic structure for crosses, nichos, and, to a lesser extent, frames.

The two large, complex frames shown in Figures 5.33 and 5.34 contain their original prints, probably French Turgis lithographs from the mid-nineteenth century. Although the prints are probably not ex-

actly the same age, the execution and design of the frames indicate that they were at least contemporary. The frames are constructed as elongated octagonals of glass and wallpaper panels connected by stamped triangular pieces of tin. The form is overlaid with the fanciful cut and stamped images that give these pieces visual strength and a delightful originality. A common template was used to create identical circles, arches, and birds, though the stamping is varied. The arches that form the perimeter of the octagon in Figure 5.33 are ornamented with a border of closely placed C stamps, and the interior contains a zigzag design created by single-line scoring on both sides of the tin piece. Discs within the arches are decorated with the same closely stamped C border surrounding a scored rosette. Fan-shaped pieces scored in a similar manner connect the rectangular glass panels. The eight birds are obviously cut from the same template, each ornamented with a series of lines created by the repetitive C stamp. The glass panels in this frame have been extensively damaged, and subsequently the wallpaper is badly faded. The poor condition of the piece does not allow for definite dating.

The frame in Figure 5.34 is in considerably better condition. It is identical in form and size to the previous one, but somewhat different in ornamentation and design. The discs have the same C border, but in place of the central rosette star shapes have been applied over the disc. The birds, also stamped with the C stamp, are indistinguishable from those in the previous example. The frame is equipped with one partial and two complete original candle holders. Five overlaid commercial mirror discs may not have been a part of the original design, but

CHARACTERISTICS OF
MORA OCTAGONAL
WORKSHOP

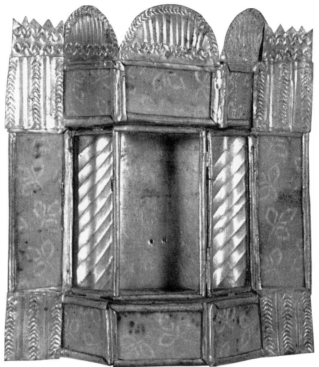

5.35

5.36

FIGURE 5.35
*Sconce, one of a pair.
Mora Octagonal
Tinsmith, circa 1870,
12³/₄ × 13¹/₂ × 3³/₄.
Southwest Museum,
Los Angeles,
California.*

FIGURE 5.36
*Nicho with cloth
under glass. Mora
Octagonal Tinsmith,
circa 1870, 10¹/₄ ×
9¹/₂ × 2. Charles D.
Carroll bequest to the
collections of the
Museum of
International Folk
Art, a unit of the
Museum of New
Mexico, Santa Fe.*

attached at a later date, possibly to reflect the glow of the candlelight.

Figure 4.29 is a tin nicho fitted to a wooden stand that features attached palm-frond forms and a wooden arch. The piece is attributed to the Arroyo Hondo Painter. Another piece attributed to a santero is the bulto of San Miguel in Figure 4.27 made by José Raphael Aragón. Cut with a small scalloped cutter and carefully stamped, the crown of the bulto includes a number of details typical of this workshop. The tin nicho also has stamping very similar to the crown. The only surviving example of a santero-painted wooden nicho with attached tin decorations is illustrated in *The Cross and the Sword* exhibition catalog.[19] Attributed to the school of José Raphael Aragón, the retablo/nicho is decorated with strips of stamped tin applied over the painted panels. It is not known at what point the strips were applied. They are unquestionably by the same hand responsible for the two octagonal frames in Figures 5.33 and 5.34.

The sconce seen in Figure 5.35 is decorated with

5.37

5.38

FIGURE 5.37
Nicho. Mora
Octagonal Tinsmith,
circa 1875, 18¹/₂ × 16
× 2. Collection of
Robert Gallegos,
Albuquerque.

FIGURE 5.38
Frame with embossed
oleograph. Mora
Octagonal Tinsmith,
circa 1890, 14¹/₂ ×
11¹/₄. Collection of
Ortíz y Pino family,
Galisteo.

the same ornamental details as the frames. The center of each sconce features the same type of rosette as can be seen in Figure 5.33 created by reverse stamping with a toothed bar stamp. The birds are similar in shape, though the wings are more deeply cut. These sconces were originally part of the collection of Felipe Delgado, gathered before 1935 presumably in the Santa Fe area.[20]

The shallow nichos seen in Figures 5.36 and 5.37, although not dramatic or exceptionally beautiful, are representative of the Mora Octagonal

Workshop. The nicho in Figure 5.36 has glass and wallpaper panels bound by tin corner pieces closely stamped with the C die. Three lunettes stamped and scored in a manner similar to that found on the connecting pieces of the frame in Figure 5.33 surmount the nicho. The piece's diagonally scored engaged columns are similar to those of the Santa Fe Federal Workshop. Mid-Victorian wallpaper panels form the body of the simple nicho in Figure 5.37. Rosettes in the four corner pieces are similar to those in the center of the sconce (Figure 5.35) from

5.39

5.40

the Southwest Museum. The border of corrugated tin strips, unusual in New Mexican tinwork, is more commonly seen in Mexican examples. Both nichos were made to contain bultos. There is no evidence to suggest that a framed print or Holy Card was ever attached to the back panel of either nicho.

A related group, differing in design and ornamentation, appears to have been executed by another Mora Octagonal tinsmith. Figures 5.38 and 5.39 are excellent examples of the more elaborate tinwork of this maker. The boldness of the pieces

is in part a result of the deeply scalloped outline created by the horseshoe-shaped cutter. Surface ornamentation is distinctive, accomplished with back-to-back stamping with a large, notched crescent stamp. The back panel of the nicho has a scalloped square opening which was cut with a small curved chisel. A print of San Antonio pasted onto delicate floral wallpaper is attached behind the opening.

A nicho with a similar detail is shown in Figure 5.40. Here, the opening frames a piece of wallpaper and a Benziger Holy Card. The tin side panels are

FIGURE 5.39
Nicho with French Holy Card. Mora Octagonal Tinsmith, circa 1880, 17³/₄ × 13 × 2⁵/₈. Taylor Museum, Colorado Springs, Colorado.

FIGURE 5.40
Nicho. Mora Octagonal Tinsmith, circa 1875, 16³/₄ × 15¹/₂ × 3¹/₄. Taylor Museum, Colorado Springs, Colorado.

5.41

5.42

FIGURE 5.41
*Cross with wallpaper.
Mora Octagonal
Tinsmith, circa 1875,
8¹/₂ × 8. Collection
of Taylor Museum,
Colorado Springs,
Colorado.*

FIGURE 5.42
*Frame with German
lithograph. Mora
Octagonal Tinsmith,
circa 1880, 14¹/₂ ×
12¹/₂. Collection of
Ortíz y Pino family,
Galisteo.*

decorated with C stamping interlocked to produce a continuous chain. Tightly controlled star-burst patterns appear on the back panel and lunette.

The simple frame in Plate 7 is extensively ornamented with the reverse C stamp. The border is stamped with a series of toothed crescent dies and embossed with a small star stamp. The simplicity of the frame complements the complex print of "Saint Veronica's Veil" and the Victorian wallpaper border strips. An identical border paper is seen in the cross in Figure 5.41. One of a pair, the cross is

formed of glass panels with deeply scalloped, stamped, and embossed quarter circles attached at the intersections. The semicircular finials bear the stamping associated with this workshop.

Figure 5.42 is similar in concept to nichos and frames that are more spare in design. The simple frame contains a lithograph of the "Sacred Heart of Jesus." Attached are two stamped and embossed side scrolls and a stamped lunette. The shape of the scrolls is similar to that of the lunette of the nicho in Figure 5.40. The familiar toothed crescent stamp is again employed as a major decorative element. These simple frames are difficult to attribute to any particular workshop. The border decoration and the placement of the rosette as well as the shapes of the side pieces, however, are similar to easily recognized pieces from this workshop.

The complex Mora Octagonal Workshop has several recognizable characteristics that link pieces together. It seems apparent that three or more craftsmen worked during a forty-year period; the resultant work displays numerous cross-influences and similarities.

José Mariá Apodaca—The Small Scallop Tinsmith

José María Apodaca (1844–1924) was a prolific tinsmith whose work combines sophisticated design, fine craftsmanship, and a superb sense of color, resulting in some of the highest quality New Mexican tinwork produced in the late nineteenth and early twentieth centuries. He was born in Paseo del Norte (Juarez, Mexico) and immigrated to Santa Fe County while still in his teens. He homesteaded land in Ojo de la Vaca in the 1880s and lived and worked for the remainder of his life in the small village (now abandoned) twenty miles southeast of Santa Fe.[21]

Mr. Apodaca was a producing tinsmith for at least forty years, from 1875 to 1915. His grandson, Ben Apodaca Martínez, remembers that he worked in the kitchen of his home surrounded by stacks of tin cans, constructing the tinwork on the kitchen

5.43

table and heating his soldering irons in the cast iron stove. Mr. Apodaca packed his tinwork in canvas bags and traveled by horseback to Santa Fe and along the Pecos River, selling his work in the villages of Colonias, San Isidro, San José, and Villanueva. Owing to his age and failing eyesight, Mr. Apodaca stopped producing tinwork about 1915.[22]

Represented by a variety of frames, nichos, boxes, and miscellaneous pieces, his work is easy to recognize. The most distinctive characteristic is extensive surface decoration produced by a small crescent-shaped cutting tool and toothed crescent-shaped stamps. The small cutter produces small scalloped edges. Thin strips of tin (embossed and stamped with scalloped stamps) visually and structurally unify the glass panels of the body and facade of his nichos and frames.

FIGURE 5.43
Portrait of José María Apodaca with his wife and daughter, circa 1910. Collection of Ray Herrera, Santa Fe.

Sophisticated and imaginative combinations of wallpapers, cloth, printed materials, and colored portions of tin containers are a benchmark of Mr. Apodaca's work. Color lithographed or varnished tin is used in the interiors of nichos, as structural tubing, and more prominently on concentric rosettes. These overlapping layers of small scalloped rosettes retain the colored labels of the original cans in combinations of blue, rose, orange-red, and acid green. The coated interior of cans (a transparent golden color) produces a dazzling effect in combination with the silver tone of the exterior. In less talented hands this technique might have seemed garish, but Mr. Apodaca produced pieces of extraordinary beauty by judicious use of this detail. Light— even ephemeral—in appearance, the work also owes much of its charm to a selective use of pastel-colored paper, seed packets, and pages from seed and nursery catalogs.

Mr. Apodaca produced two styles of frames: flat frames containing large chromos or lithographs and three-dimensional frames made with a flat outer frame surrounding a recessed inner frame that held a small Holy Card or devotional print. The main panels of frames and nichos include numerous small rectangles of patterned wallpaper, cloth, or colored pictures from seed packets protected with glass. The seams of the glass panels are covered with thin strips of stamped and embossed tin. This creates a series of smaller symmetrical panels within the larger framing panels. Pediments are made in a similar manner to match or complement the side panels.

He also produced small, simple nichos as well as large, complex ones. The multiple Gothic Revival pediments of nichos, which may be architec-

tural references, include a larger central pediment usually flanked by two smaller, more vertical pediments. Elaborate interiors often contain small framed late-Victorian lithographs soldered to the back panel as well as tin tubes and scalloped platforms made to hold paper flowers, small bultos, or other devotional items. With these additions, the interior space essentially becomes a shadow-box shrine.

Other objects produced by Mr. Apodaca include a few mirrored sconces and small oddities apparently designed to hold either thin tapers or handmade paper flowers. While no crosses from his workshop have been located, there are numerous examples of trinket boxes in collections. The late date of production (the last two decades of the nineteenth century and the first decade of the twentieth century) suggests that coal-oil lamps and other improved lighting techniques had eliminated the need for sconces other than for purely decorative purposes.

His more complex and sophisticated work is readily recognizable; however, there are a number of frames and nichos that share some of the design details but are far more simple, with less ornamentation. Some of his work was restrained and conservative, while other pieces were more creative, energetic, and experimental. The restrained work appears to be somewhat earlier, judging by the number of frames containing French and Swiss devotional prints from the last quarter of the nineteenth century. The wallpaper backing for these prints is usually a late-Victorian style that suggests a date as early as 1880. The more complex pieces feature greater numbers of turn-of-the-century wallpaper and are almost pristine in condition, indicating a later

5.44

5.45

date. They were made at about the beginning of the twentieth century and continued to appear for another fifteen years. It is reasonable to assume that the simpler and less colorful pieces are earlier and the more elaborate pieces are later.

The work, then, can be separated into two phases—the First Phase, from about 1875 to 1895, and the Second Phase, from 1895 to approximately 1915. A Byzantine quality of the design and decoration, the sure craftsmanship, and the controlled use of color and pattern are characteristic of the work of the Second Phase. The inclusion of paper and cloth flowers, milagros, and related devotional

items in the Second Phase work is more commonly seen in Mexican tinwork than in the work of any other New Mexican tinsmith. Mr. Apodaca was familiar with Mexican tinwork from his boyhood years in Juarez and brought some of these design concepts north when he came to New Mexico.

Figure 5.44, an example of the First Phase, represents the most simple form of frame. The emphasis is on the framed souvenir print of "Notre Dame de Lourdes." The side and end panels contain a delicate Victorian era wallpaper that complements the French text and accompanying steel engraving. The edges of the corner pieces are simply

FIGURE 5.44
Frame with French devotional souvenir. José María Apodaca, circa 1890, 8³/₄ × 8⁵/₈. Private collection, Santa Fe.

FIGURE 5.45
Frame with Benziger Brothers catalog page. José María Apodaca, circa 1890, 11³/₄ × 13³/₄. International Folk Art Foundation collections at the Museum of International Folk Art, a unit of the Museum of New Mexico, Santa Fe.

5.46

5.47

FIGURE 5.46
Nicho with Boasse Lebel French Holy Card. José María Apodaca, circa 1890, 12¹/₂ × 10 × 3. Collections of the Museum of International Folk Art, a unit of the Museum of New Mexico, Santa Fe.

FIGURE 5.47
Nicho with candle holders. José María Apodaca, circa 1900, 14 × 9³/₄ × 3¹/₂. Private collection, Santa Fe.

decorated with a small crescent stamp that forms a series of inverted and interlocked scallops, and a small rosette created with inverted toothed crescent stamps is centered in each corner piece.

The horizontal frame in Figure 5.45 is very similar. Framing a page from a Benziger Brothers catalog advertising chromolithographs, this piece presents a fascinating puzzle: Why were the descriptive and pricing portions of the advertisement included with the pictures? It is known that the artist was unable to read English and he simply allowed the text to form a part of the decorative composition, unaware of its commercial nature.[23] The piece shows that any readily available devotional image might appear in tin frames. The top and side panels of the frame include a subtle peach and brown wallpaper set under glass panes. The four corner pieces are ornamented with typical embossed rosette and small stamped star designs.

The nicho in Figure 5.46 represents the most obvious link between Phase One and Phase Two. The two lower corner pieces are stamped similarly

5.48

5.49

FIGURE 5.48
*Nicho-frame with
Holy Card and cloth
flowers. José María
Apodaca, circa 1895,
12 × 10¹/₂ × 1¹/₄.
A.R. Mitchell
Memorial Museum
and Gallery,
Trinidad, Colorado.*

FIGURE 5.49
*Frame with Holy
Card and seed
packets. José María
Apodaca, circa 1900,
10¹/₄ × 10.
Collections of the
Museum of
International Folk
Art, a unit of the
Museum of New
Mexico, Santa Fe.*

to the two previous frames, except that there are multilayered rosettes typical of the Second Phase at the apex of the nicho. The elongated triangular form, Gothic in character, is suggestive of the more elaborate pedimented nichos seen in Plate 8 and Figures 4.4 and 5.47. The interior contains a framed Holy Card surrounded by an embossed and stamped border depicting Jesus as a youth. The unusual shallow form of this nicho was designed to highlight the enclosed print, with no intention of constructing a space to enclose a bulto. This concept, which

is typical of most Mexican nicho-frames, shows that Mr. Apodaca was familiar with Mexican examples (see Chapter 7).

The charming piece seen in Figure 5.48 represents an excellent example of classic Phase Two. The basic form is rectangular, with side panels containing three pieces of glass that display alternating pieces of floral patterned wallpapers. Each corner piece is covered with a multilayered scalloped rosette. Edges of the glass panes are covered with stamped tin strips. This nicho-frame, which is con-

5.50

FIGURE 5.50
*Sconce with mirror,
one of pair. José
María Apodaca, circa
1900, 7-inch
diameter. Private
collection, Santa Fe.*

structed as a shadow box, contains a framed Holy Card of Jesus as a youth with a cross. The inner frame is outlined with tin strips stamped with the same repetitive serrated-scallop stamp used to ornament the outer frame. Attached to the sides of the inner frame and protected by the outer layer of glass are four paper or cloth pansies which complement the floral wallpaper of the outer frame.

Figure 4.4 is a fine example of the development of Second Phase frames. Incorporating central and flanking pediments, this design appears frequently in the late work from this workshop. Although simply constructed, an illusion of complexity is produced by the energetic patterned wallpaper used in the framing panels. These are juxtaposed against a large central panel of floral wallpaper (a backdrop for the cut-paper image of St. Anthony). Here is

another instance in which a religious catalog illustration becomes an object of veneration.

The two nichos in Figure 5.47 and Plate 8 are related to the previous frame in theme and style. Figure 5.47 shows a nicho with similar construction; here, however, the frame becomes the facade of the nicho. The center panel functions as a door ornamented with stamped and embossed lacquered tin. Semicircular platforms have been added to each side of the nicho to accept candle holders. Plate 8 is a creative and beautiful example of Mr. Apodaca's Second Phase. Its horizontal format is unusual—an excellent example of the three-dimensional theatrical quality sometimes seen in examples of this workshop. The effect is heightened by the use of clear glass panels that form the sides and tops of the nicho that are concealed behind the facade. This allows light to illuminate dramatically the image and paraphernalia in the shallow interior.

The nicho in Plate 9 is the largest and one of the most outstanding examples produced by Mr. Apodaca, the ambitious piece exemplifying his care and skill. The choice of lavender wallpapers complemented by strips of embossed and stamped gold-tone lacquered tin is inspired. The door of the nicho, constructed as a single pane of glass, is decorated with a stamped and embossed tin panel defining three arches, the central arch proportionately larger than the flanking ones. The panel is further accented with four perforated rosettes. The entire composition suggests Romanesque Revival architecture of the late nineteenth century. Perforated rosette motifs in the door panel are repeated in embossed form on the back panel, and rosettes that typically crown the pediments of other nichos have been replaced by small crosses.

This versatile craftsman also excelled in making ordinary frames and sconces. Figure 5.49 is one from a set of two hexagonal frames. The glass panels contain fruit and floral images from seed packets or nursery catalogs. The tin portions are extensively decorated with stamping and embossing, accented by delicate scalloped edges.[24]

One of a pair of sconces illustrated in Figure 5.50 is a rare example of a lighting device from this workshop. The simple round shape has been transformed by piercing the interior surface into eight equal triangles with a sharp scalloped cutter. The triangles were recurved to form an octagonal inner shape in order to expose a mirror that reflected the light of a candle. The pair of sconces is quite colorful; small tin rosettes enameled chartreuse green and cadmium red light are applied around the opening. When originally used, the sconces must have presented a delightful complement to a family altar.

Two other noteworthy items are the flower holder in Figure 4.33 and a small trinket box in Figure 4.25. The flower holder is part of the larger set discussed in Chapter 4. The trinket box (see Chapter 4) has characteristic scalloped edges, wallpaper and glass construction, and overlaid multicolored scalloped rosettes. The small heart attached to the center of the lid is a sentimental touch consistent with the character of his work.

We are fortunate that some of his descendants have carefully preserved not only examples of his work but memories of him working in his tinshop. Certainly Mr. Apodaca will be remembered for the complex originality of his designs and his pristine craftsmanship.

Rio Abajo Workshop

The Rio Abajo Workshop, named for its place of origin, produced some of the most elaborate tinwork made in New Mexico in the nineteenth century, including nichos, frames, crosses, and chandeliers. In addition to more fanciful pieces, the tinsmiths made numerous simple frames in standard sizes for devotional prints. A consistent use of closely placed, tightly controlled stamping is characteristic of this workshop. Smooth and notched crescent shapes, notched bar stamps, single-dot punches, and toothed circular stamps combine to form complex patterns (see stamping details). Scored lines are another dominant decorative device, used alone or in combination with stamping.

Numerous pieces from this workshop have been collected in the Belen–Los Lunas area of central New Mexico, and they have been found as far south as Socorro, New Mexico. The frame and nicho in Figure 5.51 and Plate 10 both were collected from a farm family in Los Chavez, Valencia County, New Mexico. Because few of these frames retain their original prints or show printed or embossed can labels, dating is difficult. Some of the surviving prints are Currier and Ives lithographs from the mid-1870s, and there are also a number of Benziger chromolithographs and oleographs from the end of the nineteenth century. With the limited information available we conjecture that the workshop produced from 1875 to 1900.

Two simple Rio Abajo Workshop frames are illustrated in Figures 5.52 and 5.53. Both are made from narrow tin strips that have been carefully stamped, embossed, and scored. They are con-

CHARACTERISTICS OF
RIO ABAJO WORKSHOP

5.51

5.52

FIGURE 5.51
*Frame. Rio Abajo
Workshop, circa 1885,
37 × 36¹/₂. Collected
in Los Chavez, New
Mexico. Collection of
Shirley and Ward
Alan Minge, Casa
San Ysidro.*

FIGURE 5.52
*Frame with catalog
page. Rio Abajo
Workshop, circa 1890,
15 × 13¹/₄. Collected
in San Miguel
County, New Mexico.
Collection of Will
and Deborah
Knappen, Santa Fe.*

nected by fan-shaped corner pieces that are scored, stamped, and have deeply scalloped edges. The prints are held in place by tin tubes attached to the back of the framing strips. This method of construction may indicate that the frames were prefabricated and assembled after the buyer had chosen the print.

The horizontal format of the frame in Figure 5.53 is uncommon in New Mexican tinwork. This is one of a pair of frames now in the collection of the Taylor Museum, Colorado Springs, Colorado, that contains Currier and Ives lithographs from the 1870s. A triangular pediment and attached rosette give additional importance to the frame. The circle of angels suspended above the enclosed casket (containing the body of San Francisco Javier [sic]) repeats the tin rosette image of the frame.

The nicho in Figure 5.54 is constructed of glass panels joined at the bottom by two scored, fan-shaped corner pieces similar to those in Figure 5.52. The ornamentation on the upper fans is more elaborate. Multipointed stars and a quilted diamond pattern have been produced with a short, toothed bar stamp,

5.53

5.54

the quilted design repeated both in the tin panel that defines a pedimented opening in the door and in the open lunette. Birds perched on either side of the lunette lend a fanciful quality. The glass side panels feature an unusual technique—rosette and spade shapes cut from pastel papers, overlaid on a dark-colored background, then carefully encased beneath the glass.

The cross in Figure 5.55 is constructed with panels containing delicate late-Victorian wallpaper. The scalloped quarter circles attached to the inter-

sections of the panels are similar to corner pieces on frames. Three crescent-shaped finials on the arms of the cross are decorated with staccato punching resembling chicken tracks, made with the toothed bar stamp. The same effect is seen in the chandelier illustrated in Figure 5.56. Though minimally stamped, the dies are the same as those in the cross.

By any standard, the frame in Figure 5.51 is one of the most imaginative and remarkable examples of New Mexican tinwork. The large frame is truly a masterwork of a very gifted craftsman. The basic

FIGURE 5.53
Frame with Currier and Ives lithograph. Rio Abajo Workshop, circa 1880, 15³/8 × 17¹/2. Taylor Museum, Colorado Springs, Colorado.

FIGURE 5.54
Nicho with cut paper under glass. Rio Abajo Workshop, circa 1885, 20¹/2 × 14¹/2 × 3¹/2. Collection of Ford Ruthling, Santa Fe.

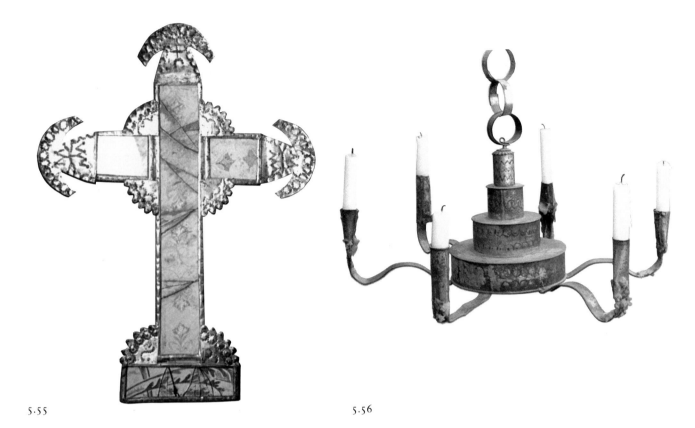

5.55

5.56

FIGURE 5.55
Cross. Rio Abajo Workshop, circa 1885, 16³/₈ × 12¹/₂. Kit Carson Foundation, Taos.

FIGURE 5.56
Chandelier. Possibly Rio Abajo Workshop, circa 1875, 8¹/₂ × 17¹/₂. Collected from a dance hall in Taos County, New Mexico. Collection of Shirley and Ward Alan Minge, Casa San Ysidro.

form consists of four side and end panels ornamented with three bands of inverted crescent stamping. These are joined by simple quarter circles outlined with more of the same stamping. A scored and stamped lunette is mounted at the top, along with two similar half-round, scored, and pierced discs at the sides. Large scored, angular swags appear below the side pieces. Two other large scored and pierced semicircles are attached to the lunette and to the half-round side pieces. The scale of these additions alone gives the frame a monumental qual-

ity rarely seen in New Mexican tinwork. The frame now contains a mirror but probably was originally made for a print. One wonders what image was so venerated that it warranted the craftsmanship to make such an elaborate frame. It is painted with a monochromatic combination of semitransparent substances resembling varnish, orange shellac, and asphaltum. Now crackled and transformed, the coated areas glow with a rich amber and brown patina in contrast to the pewter-colored surfaces of the exposed tin.[25]

By far the most ambitious piece of New Mexican tinwork is the Los Chavez nicho in Plate 10. Ornamented with an amazing variety of decorations, it is a unique example of a corner nicho. The body is trapezoidal, with tin side and skirt panels and quarter-round corner brackets. It is topped with three half-round lunettes. The borders are defined with scored lines combined with rows of closely placed, interlocking crescent stamps. Two glass side panels and the pane of glass forming the door are encased in sheets of tin, pierced with keyhole openings similar to doorways and arches found in Moorish architecture. In fact, the entire nicho has an architectural quality which is emphasized by the volume of the interior, accented with diagonally scored engaged columns. Its decorative character is augmented by the addition of forty-nine scored and stamped rosettes attached to the frame and lunettes. Multipointed stars stamped on the door panel and in the centers of the corner rosettes are similar to those seen in Figure 5.54. The six-pointed star designs in the three lunettes recall designs from the Valencia Red and Green Tinsmith of the same area of New Mexico. The use of applied rosettes around the perimeter of the lunettes is another design detail used by the Valencia Red and Green Tinsmith. Two eccentric birds flanking the lunettes (also seen in the frame of the Mathews photo) more closely resemble menacing birds of prey than doves of peace.[26]

Why were these pieces considered important enough to warrant such careful creativity in their construction? Why have only three examples with such strength and originality been seen? A simple answer may be that the cost and effort was expended for only a few extremely important commissioned pieces, or that they were for the maker's own family. Usually, the tinsmith was content to produce standardized frames and smaller nichos. So here is an artistic anomaly.

The work of the Rio Abajo Workshop is superior in style and craftsmanship to that of the Mora Octagonal Workshop, although there is some similarity between the two as well as between the Rio Abajo and the Valencia Red and Green Tinsmith. It is this cross-influence that makes the study of New Mexican tinwork both fascinating and frustrating as one attempts to distinguish the individual styles.

The Valencia Red and Green Tinsmith

A talented and prolific tinsmith working in territorial New Mexico in the late nineteenth century was responsible for the large body of work called Valencia Red and Green. Examples have been located throughout New Mexico as well as in collections in California and Colorado. It is significant, however, that numerous pieces have been collected from small villages in the Belen–Los Lunas area of Valencia County, and it is here that we believe the work originated.

The work of this craftsman can be dated circa 1870–1900 based on examples made from cans bearing patent dates or packing-company labels from the late nineteenth century. It is immediately recognizable and readily identified by its combination of dark scarlet and forest green paint on the tin surface. This group also is characterized by extensive single-dot punching and embossing used to raise patterned areas on the tin which were then usually filled with red and green pigments.

Decorative details vary from piece to piece, but there are designs which are repeated again and again on pieces of the same size and function. For example, stylized floral and leaf designs of similar size and shape appear with regularity on the end and side panels of rectangular frames. An applied rosette motif is used consistently around the perimeter of frames, on lunettes, and around the borders of nichos. Complementary colors placed against the unpainted tin surface create a strong visual tension that sets the work of this master apart from that of other tinsmiths. Though the predominant colors are red and green, blue, yellow, and light green are sometimes substituted.

The technique most often used to define the areas of pattern and design was the single-dot line, used either alone or in multiples. Often each line is reversed—one line is embossed while the adjacent line is stamped. This device is used frequently to define borders of frames and nichos.

The Valencia Red and Green Tinsmith consistently produced work which gives the impression of monumental scale. Some of the smallest works have a presence and sureness of execution equal to pieces of far greater size. The main type of work produced was frames of various sizes and shapes. Nichos, religious items such as processional crosses and torches, flat wall-hung crosses, and household items were also made.

Frames by the Valencia Red and Green Tinsmith

are generally round, rectangular, or cruciform in shape. Of fifty-four frames surveyed, thirty-six are round and are either cut from a single piece of tin or, more commonly, assembled from many small pieces of tin. Often, these multiple-part frames combine small triangular, crescent-shaped, or semicircular pieces. The development of the large form from small parts and pieces suggests that the design depended on leftover pieces of tin salvaged from remnants of larger shapes. This construction allowed a wide variety of interpretation that resulted in frames of unusual elegance and sophistication. An excellent example is a large elaborate frame in the collection of the Colorado Historical Society that is currently on exhibit in Fort Garland, Colorado.[27]

Valencia Red and Green rectangular frames vary in size from as small as 6 × 7 inches to as large as 23 1/4 × 31 1/4 inches, with the larger size more prevalent and most easily recognized. They are typically constructed of four wide, rectangular side panels accented at the corners by a rondel or elaborate rosette. Often, the frame is crowned with a lunette and also ornamented with a series of small rosettes which repeat the corner rosettes.

Two similar frames are those in Figure 5.63 and Plate 11. The first is assembled from numerous parts, resulting in a large and very elaborate frame. The second, though appearing to be made of many parts, is actually fabricated from one large sheet that has been stamped, embossed, and painted to suggest an assemblage.

The frame seen in Figure 5.57 is a good example of the variety, skill, and creative imagination evident in the craftsmanship of the Valencia Red and

CHARACTERISTICS OF THE VALENCIA RED AND GREEN TINSMITH

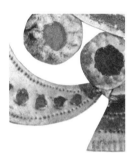

5.57

5.58

Green Tinsmith. The piece has undergone transformation through the years; most of its original paint has been lost, the original print has been replaced by a mirror, and the electrified side sconces added early in the twentieth century have been removed. The original stamping and embossing of the tin allows this frame to serve as a good example of the virtuosity and skill of the tinsmith.

The side and end panels contain examples of the reverse single-dot embossing that is characteristic of this workshop. The design of the fan leaf form is created by lines of single-dot punching, this motif appearing regularly in the central portion of each panel. In addition, a scored diaper pattern appears in the corners of each panel. The borders are ornamented with scored parallel lines, single-dot punched lines, and elaborate combinations of embossed rosettes, triangles, and reversed U stamping. The frame, which is a simple shape, lacks the typical semicircular lunette. It is one of the most complex and elaborate examples of combined techniques, however.

FIGURE 5.57
Frame. Valencia Red and Green Tinsmith, circa 1875, 22 × 10. Private collection, Santa Fe.

FIGURE 5.58
Frame. Valencia Red and Green Tinsmith, circa 1880, 20¹/₂ × 16. Private collection, Santa Fe.

5.59

5.60

FIGURE 5.59
*Frame with
oleograph. Valencia
Red and Green
Tinsmith, circa 1895,
31¹/₄ × 23¹/₄.
Collection of Will
and Deborah
Knappen, Santa Fe.*

FIGURE 5.60
*Frame with Benziger
Holy Card. Valencia
Red and Green
Tinsmith, circa 1885,
7³/₄ diameter.
Collection of Mr. and
Mrs. Murdoch
Finlayson, Santa Fe.*

While the frame in Figure 5.57 represents the complexity of technique achieved by the Valencia Red and Green Tinsmith, Figure 5.58 exemplifies the degree of elegance, fantasy, and sophistication of design which the craftsman was capable of producing. Each side panel contains the tinsmith's characteristic fan-and-leaf motif. Borders are made of a series of repeated embossed and stamped rosettes, their floral nature carefully defined and separated from the fan-and-leaf design by the simple yet effective use of the single-dot punched and em-

bossed line. Large rosettes made by embossing, stamping, and reverse single-line scoring accent the corners individually. The lunette is stamped, embossed, and made to appear three-dimensional by the use of alternate single-line scoring, which gives the appearance of an open fan. It is further enhanced by the addition of small, separately embossed rosettes which repeat the design along the borders of the side panels. The entire piece is decorated with a palette of transparent ochre, red, and green paint faded to a lovely soft patina.

5.61

5.62

Figure 5.59 is a valuable example because of its nearly perfect condition. It is made from lard cans bearing the label of the Colorado Packing and Provision Company and dating between 1892 and 1899. This frame is an example of the late work of the Valencia Red and Green Tinsmith (the earliest dated piece of tin to appear in the work is from 1872). A stamped patent date is integrated into the design of a rosette on a frame similar in concept and form, illustrated in Figure 5.57. The frame in Figure 5.59 retains its original brightness; the tin has not tar-

nished, the paint has not faded or chipped, and the chromolithograph is original. Unfortunately, there exists no collection data regarding this frame.

The Valencia Red and Green Tinsmith seems to have had a penchant for creative and unusual forms. This is evident in sconces and numerous frames containing very small devotional prints as in Figure 5.60. Related to round frames assembled from variously shaped small parts, the more creative frames also began either as a small circular shape or as a small rectangle. The sconce in Figure 5.61 and the

FIGURE 5.61
Sconce. Valencia Red and Green Tinsmith, circa 1880, 10 1/2 × 8 × 2 1/4. Collection of Will and Deborah Knappen, Santa Fe.

FIGURE 5.62
Frame, one of pair. Valencia Red and Green Tinsmith, circa 1885, 18 × 17. Collection of Ford Ruthling, Santa Fe.

5.63

5.64

round frame with mirror in Figure 5.62 are representative of this technique. From these basic shapes
evolved both cruciform and round designs with a
delicate and lacy quality. Developed from small
crescent shapes combined with triangles or rosettes, frames such as the one seen in Figure 5.63
evolved into large, elaborate works of great fragility.
Because of their intrinsically fragile nature, few of
these elaborate, imaginative works have survived
intact. Figure 5.64 is a delightful example of the
delicate quality achieved by the simple technique

of arranging stamped crescent shapes and scored
rosettes consecutively around a round or rectangular core. The addition of a cross (made of tin
tubes) to the top of the piece lends a sense of sweet-
ness and piety, and the use of color an added note
of grace.

In addition to making two-dimensional frames,
the Valencia Red and Green Tinsmith also constructed three-dimensional nichos. Among more
than seventy works surveyed from this workshop,
only a few nichos have been found. This scarcity

5.65

5.66

FIGURE 5.65
Nicho with hand-drawn Santo Niño de Atocha. Valencia Red and Green Tinsmith, circa 1880, 18¹/₄ × 14¹/₄ × 2¹/₄. Collected in San Miguel County. Collection of Will and Deborah Knappen, Santa Fe.

FIGURE 5.66
Nicho. Valencia Red and Green Tinsmith, circa 1880, 28¹/₂ × 25 × 13. Spanish Colonial Art Society, Inc. collection on loan to the Museum of International Folk Art.

seems unusual, because the strength of execution of these nichos suggests that the workshop must have created a number of them. Others must certainly exist today, lovingly tended on family altars, in homes, or in private collections.

The nicho in Figure 5.65 illustrates the degree of care, craftsmanship, and quality of design from this workshop. The piece, intact and in excellent condition, retains traces of original paint. The border is defined with a series of semiopen fans, executed by single-line scoring with single-dot embossing and an unusual meander design that defines the inner borders of the side and end pieces as well as the lunette. Three scored and fluted arrow-shaped finials hint at Gothic Revival influence. The sophisticated workmanship and excellent design suggest the maturity of craftsmanship that appears after long years of work.

Figure 5.66 is an example of a larger nicho less successful in either design or scale. It is constructed of wide side panels that are scored, stamped, and embossed in a manner typical of frames. The nicho

5.67

5.68

FIGURE 5.67
Detail of processional cross. Valencia Red and Green Tinsmith, circa 1880, 81 × 16¹/₂ × 1¹/₂ overall. Collection of Shirley and Ward Alan Minge, Casa San Ysidro.

FIGURE 5.68
Nicho. Valencia Red and Green Tinsmith, circa 1875, 29 × 24 × 6¹/₂. Heard Museum, Phoenix, Arizona.

is trapezoidal in cross-section and was designed to accommodate a large bulto. Diagonally scored half-columns embellish the interior. The top is crowned with an assortment of crescents, finials, and a strange rooster-like bird, and the sides are ornamented with an abstracted leaf shape identical to those on the processional cross from the mission church at Manzano, New Mexico (Figure 5.67). The size and similarity of various elements in the piece suggest that they were made at the same time. The nicho is intact though museum accession records show it

has been strengthened and the candle holders repaired. It exemplifies the use of candles to illuminate and venerate the bulto inside the nicho.

The fanciful nicho illustrated in Figure 5.68, now damaged, illustrates the use of the rectangular frame (Figure 5.57) as the basis for the construction of a nicho. With the addition of the elaborate lunette, the nicho has been transformed into an object of sophisticated design and beauty.

A rare set of three pieces made for a church is illustrated in Figure 4.31. The group consists of a processional cross with two matching torches equipped to hold candles. The pieces are constructed entirely of tin cans over a wooden armature. Enough traces of original paint remain to suggest the earlier appearance. The processional set was collected from the mission church of Manzano, New Mexico, a small village in Torrance County. Manzano lies directly across the mountains east of Tomé, and in the nineteenth century was included in Valencia County.

These pieces are the largest examples of New Mexican tinwork to survive. The cross (Figure 5.67) is three-dimensional with matching front and back panels ornamented with single-dot embossing. Each juncture of the cross is accented with a leaf form, but two lower leaf appendages are missing. The processional torches are constructed as simple funnels attached to long wooden poles covered with tin, the funnel and poles decorated with scored spirals. The torch handles show fragments of stamping that read "pure lard." The stamped portions have been incorporated into the spiral design of the handles in such a manner that the labeling nearly disappears.

Figure 5.69 is a cross of exceptional beauty in both design and craftsmanship. The design is based on alternate bands of red and green paint under panes of glass. Four right angles formed by the Latin cross each contain a quarter circle of tin which is ornamented with a scored diaper pattern and bordered with small applied rosettes. The four ends of the cross are accented with tin finials shaped to suggest a bishop's miter. The outline of these shapes is accented with single-dot embossing and bar stamping. The simple beauty and sophistication of

5.69

design of this cross makes it one of the best examples of nineteenth-century tin work in this particular genre.

Figure 5.70 is one of a pair of wall sconces. The pair retain their red and green paint and show little or no repair. In designing the sconces the maker chose a form for the background which would be large enough to accommodate safely the weight and mass of three candles without sacrificing function or beauty. The ornamentation of the pedimented tin back plate is simple line scoring and repetitive

FIGURE 5.69
Cross. Valencia Red and Green Tinsmith, circa 1880, 33 3/8 × 26. Taylor Museum, Colorado Springs, Colorado.

Valencia Red and Green II Workshop

A group of pieces that appear with regularity in collections in Colorado and New Mexico has continuously puzzled us. The work at first glance appears identical to that of the Valencia Red and Green Tinsmith. Upon closer examination, however, each piece has an individuality and a certain crudeness that prompted attribution to another workshop—the Valencia Red and Green II. Based on both design and collection information, we believe this body of work was produced by at least two craftsmen working over a substantial period of time in Valencia County, New Mexico. Dating the work is difficult. Few frames have survived with the original prints, but examples retaining their prints, and one example backed with portions of a hatbox from a firm in Albuquerque, show that the work is generally later than that of the Valencia Red and Green Tinsmith. The production period of the Valencia Red and Green II Workshop is from about 1885 to 1910. Variations of Valencia Red and Green II appeared in the work of the tinsmiths of the Revival period and are in use by some tinsmiths working today in New Mexico. Though simple in concept and crude in construction, the work of the Valencia Red and Green II possesses a strong visual imagery that imbues the work with naive charm.

Closely allied to the work of the Valencia Red and Green Tinsmith, the majority of these pieces are painted with a medium-value red and forest green oil-based paint. In contrast to the restrained and careful painting of the Valencia Red and Green Tin-

5.70

FIGURE 5.70
Sconce, one of pair. Valencia Red and Green Tinsmith, circa 1880, 11¹/₂ × 10¹/₈ × 7³/₈. Collections of the Museum of International Folk Art, a unit of the Museum of New Mexico, Santa Fe.

rows of alternate single-dot punching and embossing. Each candle holder and bobeche is supported by a thin wire which is attached to the decorative tin backing and bent into an elegant sweeping curve. There are only a few surviving sconces from this workshop.

With only a few tools and an extensive ornamental design vocabulary, the Valencia Red and Green Tinsmith created a massive body of work which has been widely appreciated and avidly collected. He continues to be cherished as one of the outstanding artists in this genre.

5.71

5.72

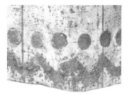

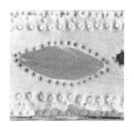

FIGURE 5.71
Frame with embossed oleograph. Valencia Red and Green II Workshop, circa 1890, 12 × 8. Collection of Ford Ruthling, Santa Fe.

FIGURE 5.72
Nicho containing bulto of Santo Niño de Atocha and salvaged etched glass. Valencia Red and Green II Workshop, circa 1890, 20 × 16 × 3. Collection of Mr. and Mrs. Gerald Peters, Santa Fe.

smith, the paint on this work is applied in a free and almost careless manner, indicative of a craftsman with an indifferent approach to his craft. Not only the painting but the overall construction of the work is haphazard. Undoubtedly imitating the work of the Valencia Red and Green Tinsmith, the work shares similar scale and design concepts, but the differences are quite apparent. There is less use of ornamentation and a minimal use of complex stamping. The principal method of decoration is single-dot embossing. The prevalent fleur-de-lis or fan-and-leaf designs of the Valencia Red and Green Tinsmith also appear commonly in the work of the Valencia Red and Green II, but with neither precision nor elegance. Often, frames and nichos are crowned with an odd combination of rounded leaf or feather shapes and squat, stylized birds. The birds frequently are painted half red, half green. One of the most typical and easily recognized details is the use of one-quarter- to one-half-inch circles created by single-dot embossing. The circles are filled alternately with thick red and green paint.

5.73

5.74

FIGURE 5.73
*Frame containing
Mexican Holy Card.
Valencia Red and
Green Workshop,
circa 1890, 7³/₄-inch
diameter. Collection
of Will and Deborah
Knappen, Santa Fe.*

FIGURE 5.74
*Nicho. Valencia Red
and Green II
Workshop, circa 1885,
22 × 13³/₄ × 3¹/₄.
A.R. Mitchell
Memorial Museum
and Gallery,
Trinidad, Colorado.*

The use of architectural forms in both two- and three-dimensional constructions is a unique aspect of Valencia Red and Green II pieces. Several frames by this workshop closely resemble twin-towered facades of New Mexican Spanish Colonial churches.[28] The nichos also echo the symmetrical mass of early New Mexican missions (Figures 5.71 and 5.72). Close in concept to the architectural pieces but more common in form is the charming round frame containing a Holy Card which is illustrated in Figure 5.73. A similar ornamental vocabulary

employed in the design of the three pieces suggests that they were created within a short time period.

Figure 5.74 is one of two nichos which illustrate the similarity between Valencia Red and Green and Valencia Red and Green II and also, on close examination, the obvious dissimilarities. Like Valencia Red and Green work, side panels in these pieces are ornamented with single-dot embossing in a design closely related to that in Figure 5.59. The corner rosettes and lunette are also similar in form to work of the Valencia Red and Green Tinsmith. In

5.75

5.76

FIGURE 5.75
Frame with Currier and Ives lithograph. Valencia Red and Green II Workshop, circa 1885, 29¹/₂ × 18³/₈. Harwood Foundation Museum, Taos.

FIGURE 5.76
Frame with mirror. Valencia Red and Green II Workshop, circa 1885, 45¹/₂ × 33. Collections of the Museum of International Folk Art, a unit of the Museum of New Mexico, Santa Fe.

contrast to Valencia Red and Green, the side panels are crudely cut, the border shows carelessly executed repetitive bar stamping, and the fan-and-leaf design of the panels is randomly embossed in an unrefined manner. The rosettes are stamped with the same single-bar tool used on the panels, and the lunette is simply stamped and single-dot embossed to form swags painted in typical alternating red and green. The crowning set of birds and the splayed leaf forms are distinctive characteristics. The entire piece is carelessly painted in an unusual variation

of red and green—in this instance lime green contrasted with a deep alizarin crimson that is almost maroon. This is far afield from the pure Valencia Red and Green.

Figure 5.75 illustrates a frame that closely resembles the work of the Valencia Red and Green Tinsmith. The familiar fleur-de-lis design, the stamped and embossed border, and the lunette are all typical of the earlier workshop, but the differences are also apparent. The serrated lunette is elongated in a mannered style and ornamented with

5.77

5.78

FIGURE 5.77
*Frame. Valencia Red
and Green II
Workshop, circa 1885,
26 × 17. Collection
of Ford Ruthling,
Santa Fe.*

FIGURE 5.78
*Frame. Valencia Red
and Green II
Workshop, circa 1890,
24 × 13 1/2. A.R.
Mitchell Memorial
Museum and Gallery,
Trinidad, Colorado.*

the single-dot swag design typical of Valencia Red and Green II. The corners are serrated quarter circles painted red and green without an embossed line to separate the colors. This detail does not appear in the Valencia Red and Green master's work.

A frame that has a similar serrated profile but is more elaborate and complex is presently in the collection of the Museum of International Folk Art in Santa Fe. Without its original print and fitted with a mirror, it still is one of the most ambitious pieces by this workshop (Figure 5.76). The frame is

constructed of individual parts in the same technique used by the Valencia Red and Green Tinsmith. While all parts have been simplified, they lack the sophistication of design and craftsmanship of the other work. The crest, composed of crescent, leaf, and bird shapes, is at once wonderfully whimsical and strong. An identical frame was illustrated in the 1937 state vocational workbook.[29]

Figure 5.77 also shows a good example of the more ambitious work of the Valencia Red and Green II Workshop. This frame illustrates the tinsmith's

5.79

5.80

capacity to build large-scale frames. The original print has been replaced with a mirror. The side panels are simply ornamented with single-dot embossing that creates a design like twisted ribbons which are further defined with lines of single-dot embossing that meander along the edges. Corner rosettes contain a six-pointed star and are more carefully executed than many other examples from this workshop. Crescent-shaped pieces along the sides and bottom and finials at the top of the frame lend additional form and importance. Francisco Delgado, a noted Revival period tinsmith from Santa Fe, prominently displayed either this or an identical frame in his shop on Canyon Road, circa 1935 (Figure 6.6).

Figure 5.78 represents an eccentric design from the same workshop. Made of four separate side panels similar to those of the scalloped work from the Rio Arriba Workshop, this frame features a large, simple lunette accented with five large stars. The freedom of expression created by large shapes and freely painted surfaces brings energy and freshness to an

FIGURE 5.79
Nicho. Valencia Red and Green II Workshop, circa 1885, 9¼ × 8½ × 2. Private collection, Belen.

FIGURE 5.80
Sconce, one of pair. Valencia Red and Green II Workshop, circa 1885, 10 × 9 × 2. Collected in northern Socorro County, New Mexico. Private collection, Belen.

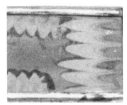

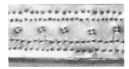

object which might otherwise be considered crudely proportioned.

The nicho and sconce (one of a set of four) in Figures 5.79 and 5.80 are perhaps the most well-preserved and classic examples of the Valencia Red and Green II Workshop. Collected in northern Socorro County, they combine the strong sense of form with the simple vocabulary of decoration so typical of this body of work. The black and orange paint decorating the surface is still bright, which provides an indication of how other work might have appeared at the time of its fabrication. These examples show the definite stylistic similarity of the products of this workshop to those of the Valencia Red and Green Tinsmith, tempered by a decidedly conservative sensibility.

The identity of the Valencia Red and Green II Workshop continues to pose questions. Why are the designs and forms so similar to those of the Valencia Red and Green Tinsmith? Is this the work of later tinsmiths trained by the Red and Green master? Or is it the work of less skilled local craftsmen adapting familiar forms into naive versions of their own?

The Isleta Tinsmith

Perhaps the most easily recognized New Mexican tinwork was made by the Isleta Tinsmith, circa 1885 to 1920. This workshop spanned the period from the tinwork styles of the nineteenth century to the beginning of the Revival period but was not a part of the Revival. It is not known whether Isleta tinwork was made by an Indian craftsman at Isleta Pueblo or a Spanish craftsman in one of the surrounding communities. Few clues are available to determine a specific location of the workshop or the identity of the maker. It is important to examine the evidence that points in both directions.

The name *Isleta Tinsmith* was chosen because numerous examples have been collected from families at Isleta Pueblo during the last fifty years. No tinsmiths were reported at the pueblo in the United States census from 1880 to 1910, and none appear in state directories of the period. Nevertheless, evidence that Isleta was the center of the area of production is overwhelming. John Adair reported that an Indian silversmith, José Padilla, who worked at Isleta in the 1870s, "learned silversmithing by seeing a Mexican, in El Paso, Texas, do some tinwork, making frames for Santos."[30] Tinwork objects were very popular in the pueblo. Four published photographs taken at Isleta by Charles Lummis between 1890 and 1900 show tinwork sconces, frames, and nichos from several of the Rio Abajo area workshops.[31] Photographs of the interior of Isleta Governor Lente's home, circa 1915, display several tin frames (see frontispiece). At least fourteen frames by the Isleta Tinsmith, most containing photographs of pueblo residents, have been located (see Figures 5.81 and 5.82). Some of these photographs were taken about 1910–15 by an itinerant tent photographer, probably during the annual Feast Day fair. This fair, which has been held for more than a century, attracted vendors and buyers from neighboring Spanish communities such as Peralta and Valencia and as far away as Albuquerque.

Three frames have been found made from salvaged cans stenciled "A.G. Seis, Isleta" (Figure 5.83), which is compelling evidence that the tinwork was

5.81

5.82

FIGURE 5.81
*Frame with
photograph of Isleta
Pueblo, unidentified
photographer. The
Isleta Tinsmith, circa
1900, 11³/₄ × 9¹/₄.
Collection of Shirley
and Ward Alan
Minge, Casa San
Ysidro.*

FIGURE 5.82
*Frame containing
three photographs,
photographer
unknown. The Isleta
Tinsmith, circa 1915,
7 × 12. Millicent
Rogers Museum,
Taos.*

made at the pueblo.³² Certainly, the tradition of silversmithing in Isleta suggests that the techniques were present. A number of older residents who were questioned do not remember a tinsmith, though many of them still display tin frames in their homes.³³

There are also connections between Isleta tinwork and Spanish villages nearby. New Mexican tinwork was collected in the village of Valencia more than fifty years ago, while Hispanic silversmiths have been reported from the small town of

Peralta, about seven miles south of the pueblo.³⁴ An Isleta-style frame has been located which has a religious print overlaid on a letterhead from Jake Levy's bargain store (La Tienda Barate) in Los Lunas. A farming community about eight miles south of Isleta, Los Lentes has long been an outpost of the pueblo, and a number of Indian families have intermarried with Hispanic families.

The strongest evidence that suggests a Hispanic origin for the work is based on the first-hand recollection of an elderly Isleta resident. He remem-

5.83

FIGURE 5.83
*Detail of lunette
stenciled "A.G. Seis,
Isleta, N.M.." The
Isleta Tinsmith, circa
1915, 25 × 29¹/₂
overall. Private
collection, Santa Fe.*

Tinsmith, and certainly the painted glass panels and other decorative features imply a familiarity with nineteenth-century Mexican tinwork. It is quite possible that Francisco Quiroz was the Isleta Tinsmith. The use of the frames for Catholic images suggests that the tinsmith had a larger market among the Spanish people of the Rio Abajo rather than solely at the pueblo. Certainly, future research will confirm the name of this tinsmith who produced work as late as the 1920s.

The large body of work attributed to the Isleta Tinsmith is composed almost exclusively of frames containing oleographs, chromolithographs, late-Victorian greeting cards, Holy Cards, and photographs. Only two small nichos (one collected at Isleta) have been located. Most frames include reverse-painted glass panels and decorative tin details. The glass was painted in a manner unique to Isleta: thin oil paint was freely brushed on, then a broad notched tool was swept across the freshly painted surface in a series of loops and scallops. After drying, a second color was applied in broad strokes over the combed paint. The glass was then backed with white or colored paper, or more often with the bright lacquered surface of the interior of a can. The rich golden color glowing through the combed surface is particularly effective in combination with the painted color. Though red and green are often detailed with metallic gold as the predominant color scheme, combinations of light and dark blue, white, yellow, pink, lavender, and burgundy also appear. With a few exceptions, the large frames have interior dimensions that match Benziger oleographs or Louis Prang chromolithographs. The standard size and prefabricated parts assembled later

bers a Hispanic tinsmith (about 1915) who rented a house at the pueblo for a month each year in order to produce tin frames for the community. He also reported that the cost of these frames ranged from 50 cents to $2.50.[35] This would explain the presence of frames made from cans shipped to the General Store at Isleta and also support the conclusion that the Isleta Tinsmith was Hispanic from a nearby village. The census for 1900 lists Francisco Quiroz working as a tinsmith in Belen, about eighteen miles south of Isleta. Mr. Quiroz was born in Mexico in 1868 and moved to New Mexico in 1885. These dates coincide with the working period of the Isleta

around a chosen print suggest a production-line approach to the crafting of these frames.

The work of the Isleta Tinsmith is somewhat more easy to date than that of other workshops. Isleta frames usually contain oleographs that were available well into the first quarter of the twentieth century. Lithographs appeared in Isleta frames as early as 1880, along with a variety of late-Victorian religious and greeting cards. Numerous frames with tin backing panels showing partial or complete can labels still existed. Most of the printed labels refer to the F.D.A. Act of 1906. Products of the Colorado Packing and Provision Company in Denver are quite common and date these pieces after 1893 (see Plate 1). A glass-paneled frame containing an oleograph of St. Patrick was recently broken; prior to its restoration, part of a newspaper page used to back the print was discovered. The fragment from the *Denver News* published in the spring of 1915 suggests a period of active production for the Isleta Tinsmith from 1895 to 1920. The ending date of 1920 is based on photographs of people dressed in styles worn during World War I.

Figures 5.81 and 5.82 show the simplest type of Isleta frame. Specifically made for the photographs, these frames feature brilliant yellow, blue, or red painted and combed glass panels joined at the corners with simple tin fans. The frame in Figure 5.81 contains a photograph of Isleta Pueblo from the turn of the century. This photograph is set within a border, while the horizontal frame with three photographs is not as carefully composed. The uncommon horizontal format was used either as a concession to popular late-Victorian taste or in response to the three images.

5.84

Another horizontal frame, illustrated in Figure 5.84, was designed to hold eight late-Victorian cards illustrating scenes from the life of Christ. The frame is made more elaborate with a lunette and side scrolls. Visible in the lunette are the machine-embossed ridges from the bottom of a large tin can and scored and embossed lines that form the image of a triangular pediment. Side scrolls are a repetitive element related in style to the earlier work of the Santa Fe Federal Workshop, though this is probably a coincidence. A similar but vertical frame is illustrated in Plate 12. In this instance the side scrolls have been simplified and the surface decoration

FIGURE 5.84
Frame containing Holy Cards. The Isleta Tinsmith, circa 1905, 16 × 18 1/2. Collection of Ford Ruthling, Santa Fe.

5.85

5.86

made more elaborate and varied. Dark blue and green reverse-painted glass complements the print of St. Augustine.

Two large frames with colorful painted glass panels exhibit some of the most decorative and elaborate details in Isleta tinwork. The red and gold frame (Plate 1) has lost its original print, but the well-preserved tin backing panel provides us with a label that is readily dated. The empty frame has an abstract effect which allows it to be appreciated independent of a print. Plate 13 is an example of

the tinsmith's excellent color sense. Stamping and embossing of the tin surfaces and the serrated pendant leaf forms are typical of Isleta tinwork. The perimeter of the lunettes on both examples is identical, which supports the idea that the tinsmith worked from a set of standard templates. Though the shapes are the same, the stamping is always varied slightly. The center of the lunette in Plate 13 has been removed by snipping out the tin, resulting in a serrated sunburst image.

Although the tin and painted-glass style characterizes the bulk of surviving Isleta frames, there are frames made entirely of tin. Two frames (Figures 5.85 and 5.86) are typical. Both are made from strips of tin can joined by projecting fan-shaped corner pieces. The first frame has a semicircular lunette decorated with a scored pediment and side scrolls nearly identical to those in Figure 5.84. The lunette is embossed with the typical rosette and four-dot stamp. A hand-colored lithograph of "Medalla Milagrosa" suggests that the frame was made circa 1885–90. The square frame in Figure 5.86 is the only known Isleta tin frame without a lunette. It is also the only known example with a print bordered by wallpaper. In most instances the print, photograph, or greeting card fills the entire opening.

Nineteenth-century prints in frames made entirely of tin suggest that this style may have preceded the painted-glass styles, although some all-tin frames contain early twentieth-century Benziger oleographs. It is likely that both frame styles were made throughout the productive career of the artist. Erratic stamping and embossing and careless craftsmanship are more apparent in all-tin frames than in glass-paneled frames, particularly noticeable in the uneven lines of single-dot embossing that parallel the edges. In spite of the craftsmanship the all-tin-style Isleta frames are always expressive and fanciful.

The frame in Figure 5.87 is an anomaly. The rectangular form is composed primarily of tin tubes that bind the St. Louis Gonzaga oleograph and the glass together. The sole ornamentation is the characteristic Isleta lunette and eccentric whimsical side pieces that contain the typical embossed patterns. A similar frame is included in the Mathews pho-

5.87

tographs from the 1930s, and three other frames, collected at Isleta, are known.[36]

Isleta tinwork, though very limited in its variety of forms and functions, is important for the quantity of large, elaborate frames that have survived. It also includes the greatest number of painted glass pieces. This workshop represents the finale of a relatively short, dynamic period of artistic creativity unique in nineteenth-century New Mexico.

FIGURE 5.87
Frame with Benziger oleograph. The Isleta Tinsmith, circa 1900, 22 × 18. Private collection, Santa Fe.

CHARACTERISTICS OF
THE MESILLA COMBED
PAINT TINSMITH

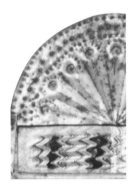

FIGURE 5.88
*Nicho-frame. Mesilla
Combed Paint
Tinsmith, circa 1910,
9³/₄ × 8. Private
collection, Santa Fe.*

5.88

The Mesilla Combed Paint Tinsmith

This body of work was named for its origin in the Mesilla Valley and for its most prominent decorative feature. It is the only recognizable style of tin work that seems to have originated in southern New Mexico. Identifying the place of origin of this tinwork is difficult. The Fred Harvey Company employed "scouts" all over New Mexico (including the Mesilla Valley) in the first half of the twentieth century. These buyers sent pieces back to Albuquerque and Santa Fe, where they were subsequently sold to both local collectors and tourists traveling cross-country.

The Fred Harvey buyers (along with buyers for J.C. Candelario, a Santa Fe curio dealer) are primarily responsible for most of the southern New Mexican pieces that appear in old collections in Santa Fe and Taos. (They are also responsible for the loss of the original collection information.[37]) Although only a limited number of pieces have been found, they were unquestionably made by a single tinsmith working in the vicinity of Mesilla in Doña Ana County, New Mexico (north of El Paso and the Mexican border). The largest single group of these pieces is in the Gadsden Museum in Mesilla.

Mesilla Combed Paint work is easily recognized by its major design element—reverse-painted and combed glass panels, usually painted with a dark brown or black oil-based paint, possibly a type of asphaltum. When diluted and applied to the glass with a dry-brush technique, the result is a rich black and amber linear design. Colored paper or metallic foil placed behind the painted glass produces an illusion of depth. Painted glass is used for side and end panels of frames and nichos, with the tin parts limited to lunettes, corner pieces, or side scrolls. The lunette is usually a scored fan, a motif repeated in the quarter-circle corner pieces. The pendant swags, or scrolls, that ornament the sides of the frames and nichos are designed as partially folded fans that terminate with an integral stamped and embossed rosette. These pieces are made from a variety of tin cans and terneplate, some bearing the

5.89

5.90

FIGURE 5.89
*Nicho. Mesilla
Combed Paint
Tinsmith, circa 1910,
21 1/2 × 17 3/4 × 3 3/4.
Albuquerque
Museum,
Albuquerque.*

FIGURE 5.90
*Frame. Mesilla
Combed Paint
Tinsmith, circa 1915,
19 3/4 × 15 3/4. Spanish
Colonial Art Society,
Inc. collection on
loan to the Museum
of International Folk
Art.*

labels of American manufacturers. Although they are difficult to date, the presence of twentieth-century terneplate and the bright condition of the tin suggests a period of production from 1890 to 1920.

The shallow nicho in Figure 5.88 was the first example located from this workshop. The work, especially the typically Mexican lunette and pendant fans, appears to be of Mexican origin. The interior of the nicho, however, retains a fragment of a label from a tin can manufactured by an American baking-powder company. The subsequent discovery of additional examples by the same tinsmith, both in New Mexico and in Arizona, implies that these pieces were made in New Mexico. Similar in design, the larger nicho in Figure 5.89 is constructed of glass panels edged with thin tubes of tin with the standard trapezoidal cross section. The top glass panels are crowned with three tin lunettes, while the side panels bear pendant scrolls nearly identical to those of the previous example. The clear glass panes that form the body are decorated with deli-

5.91

5.92

FIGURE 5.91
*Nicho-frame. Mesilla
Combed Paint
Tinsmith, circa 1900,
12¹/₄ × 10¹/₂ × 1.
Collection of Mr. and
Mrs. J. Paul Taylor,
Mesilla.*

FIGURE 5.92
*Cross with Mexican
woodcut. Mesilla
Combed Paint
Tinsmith, circa 1910,
13 × 11. Collection
of Fred McAninch,
Tucson, Arizona.*

cately painted floral designs, while the painted side panels feature a combed chevron pattern. The tinsmith has chosen to use the varnished interior of the can on the surface facade, a practice not commonly seen in New Mexican tinwork.

The example in Figure 5.90 is a framed mirror that appears to be original, not a replacement for a print. Most of the frame is fabricated from tin cans, but the backing plate is a sheet of terneplate embossed "Follansbee steel."[38] Though the design of the lunette is comparable to others, the side scrolls

are a variation of the usual pendant form. The glass panels are ornamented with parallel lines of unusual squiggle-combed paint. A similar shallow nicho (or deep frame) shown in Figure 5.91 is one of a pair collected in Mesilla. The painting on the glass is in the same style, backed with pieces of coral and yellow paper. The stamped and scored lunette is typical, but the side scrolls have become minimal and abstracted. The Mexican lithograph of San Antonio de Padua is original to the piece.

Another example from this workshop is the

beautiful cross in Figure 5.92. Chevron-patterned painted glass panels are joined with four scalloped and scored quarter circles that radiate from the intersection of the cross. A variation of the familiar lunette appears at the three ends of the cross, while the bottom features an unusual truncated triangular glass panel that gives a sense of stability to the cross. This supporting panel contains a Mexican woodcut of Nuestra Señora de San Juan de los Lagos from a popular Mexican shrine.

The strong connection of these pieces with Mexican tinwork is a riddle. A Mesilla Combed Paint nicho, from the Fred Harvey collection now at the Heard Museum in Phoenix, Arizona, is adorned with back-painted flowered glass panels similar to Mexican painted work. This nicho provides a close link with nineteenth-century Mexican pieces (see Chapter 7). The Mesilla Combed Paint tinsmith might have been a "border" tinsmith. The cross was collected in Duncan, Arizona, not far from Lordsburg, New Mexico, and the Mexican border. A long collection history in Mesilla suggests that the tinsmith could have been from Juarez, Mexico, only forty miles away. This "border" craftsman may have traveled extensively along the border from El Paso, Texas, to Nogales, Arizona, wherever there was a demand for his work; however, this theory cannot be confirmed.

A remarkable relationship exists between Mesilla Combed Paint pieces and the work of a group of tinsmiths still producing crosses, frames, nichos, and trinket boxes in the towns of Magdalena de Kino and Imuris, Sonora, Mexico, thirty to sixty miles south of Nogales, Arizona.[39] The similarities between this contemporary Mexican tinwork and the work of the Mesilla Combed Paint Tinsmith are startling. The present-day work features back-painted glass panels edged in tin, with floral or wavy-line designs. The wavy lines are coarser than those in Mesilla work, and the line-work composition is more random. Not unlike the painting on the Mesilla example at the Heard Museum, the floral painting usually includes roses and rosebuds as well as generalized leaves and stems.[40] All the painting is backed with crumpled metallic foil rather than colored paper. The crosses made by only one craftsman, Jesús Leon, are closely related to the Mesilla cross in Figure 5.92. The glass panels of the arms of the cross are supported by a similar glass-paneled trapezoidal base that also frames a commercial print. The modern Mexican crosses lack the tin finials and central rays of the Mesilla example but are supported on the back by identical tubing braces.[41]

Parallels between the two strengthen the premise that an itinerant "border" tinsmith worked extensively on both sides of the U.S./Mexican border at the turn of the century. Little is known of the history or tradition of Sonoran tinwork, but the similarity to Mesilla pieces seems more than a random coincidence.

The Fan Lunette Tinsmith

These enigmatic pieces are identifiable by their most outstanding characteristic—a crest in the form of a partially open fan. The place of origin of the work has not been determined. All of the work located thus far consist of frames, constructed of thin tubes of tin holding the prints in place under glass

CHARACTERISTICS OF
THE FAN LUNETTE
TINSMITH

5.93

5.94

FIGURE 5.93
*Frame with Benziger
lithograph. Fan
Lunette Tinsmith,
circa 1885, 15³/₄ ×
11¹/₄. Collections of
the Museum of
International Folk
Art, a unit of the
Museum of New
Mexico, Santa Fe.*

FIGURE 5.94
*Frame with hand-
tinted lithograph. Fan
Lunette Tinsmith,
circa 1870, 15¹/₄ ×
11¹/₄. Private
collection, Santa Fe.*

and without side or end panels. In addition to the crest they always include corner brackets in either an L or a modified ear shape. The flimsy construction and the absence of protective side panels have made this type of frame vulnerable to damage. Most of the frames are without their original prints, and are either empty or contain mirrors. Four frames still fitted with European prints from 1860 to 1890 have been located, however. Dating is difficult because the frames retain neither can labels nor markings. Based on the scanty evidence available, the

period of production was probably 1870–90.

All Fan Lunette frames have some stylistic details related to the work of other New Mexican tinsmiths, but we have been reluctant to assign this style to any other workshop. The form probably evolved independently. Some evidence suggests that the "ear" bracket frames were made in the Rio Abajo and are related to early work of the Valencia Red and Green Tinsmith. An 1899 photograph of the interior of the church at Isleta Pueblo shows at least four frames with similar brackets.[42]

5.95

5.96

The two frames seen in Figures 5.93 and 5.94 are obviously the work of one craftsman. Both are the same shape and almost identical in size. The tinsmith may have used a set of templates to produce identical parts for standard-sized prints. Though identical in form, the ornamentation and stamping vary slightly. The frame in Figure 5.95 still retains its fan-shaped lunette, but the L-shaped corner pieces have been replaced by rounded ear-shaped pieces of tin. Both previous examples were ornamented with stamping vaguely similar to that of the Rio Arriba

Workshop, while this example is more suggestive of the Mora Octagonal Tinsmith.

The charming small frame in Figure 5.96 may have been made by another tinsmith familiar with this style. In this example, ear-shaped corner pieces have been scored to radiate from the corners of the frame, and the fan lunette is proportionately larger than in other examples. The fan also seems to radiate from the frame—a dramatic device that draws attention to the print. The image is a delightful assemblage of embossed foil, wallpaper, and cut pa-

FIGURE 5.95
Frame. Fan Lunette Tinsmith, circa 1885, 15³/₄ × 11¹/₂. Collection of Ortíz y Pino family, Galisteo.

FIGURE 5.96
Frame with Holy Card and cut foil and tissue. Fan Lunette Tinsmith, circa 1890, 13 × 8³/₈. Albuquerque Museum, gift of Kennard F. Hertford.

per. The presence of embossed single-dot punching and circular stamping around the edges of both fan and lunette implies a relationship to the Valencia Red and Green II Workshop.

The enigmatic Fan Lunette pieces represent a unique style of New Mexican tinwork that must be separated from that of the other workshops. Their origin remains a matter of speculation, but as additional examples are found, a greater quantity of similar work will predicate a more specific attribution.

H. V. Gonzales

Without a doubt, the most frustrating aspect of our study of New Mexican tinwork has been the absence of signed pieces and the difficulty in assigning specific places of origin. Late in our research, however, an incredibly well-preserved nicho was brought to our attention that bore not only the name of the tinsmith, H.V. Gonzales, but also a dedication, the identity of the donor, and the date and place of origin. Higinio Gonzales was born in New Mexico in 1842 and was listed in the 1870 census as a tinner living in San Ildefonso. By 1880 he was no longer located in northern Santa Fe County.

The nicho (see Figures 5.97 and 5.98) is fabricated almost entirely of glass in a typical lozenge shape and, with its pediment and glass pitched roof, is related stylistically to other neogothic style nichos made in New Mexico. While in many Gothic Revival nichos the clear glass roof functions as a clerestory to illuminate a bulto or devotional print, this nicho has opaque roof panels made of painted glass backed with black oilcloth, the glass displaying the dedication and signature. Figure 5.97 illustrates both of the roof panels. The right panel contains a stylized drawing of a rose with scrolled foliage painted in yellow, pink, and green oil paint above the lettering of the dedication. In English it reads, "At the request and expense of Faustin Vigil a devoted servant of the most blessed Virgin."[43] The left panel, containing the signature, date, and location of origin, is painted with an identical rose and scroll image and can be translated to read, "Dedicated to the honor of our most blessed Guadelupe of Mexico, today the 28th day of January of 1872 by H.V. Gonzales, San Ildefonso, New Mexico." The panel of the pediment is back-painted similarly with yellow line work over a combination of the oilcloth and a blue painted ground (see Figure 5.98). The image is of two crudely drawn angels holding a long scroll above a standing figure, suggesting an image of God the Father giving a blessing.

Most of the tin surface of the nicho remains bright, and all of the glass is intact. Its present condition gives an excellent sense of the appearance of newly made New Mexican tinwork. The back of the nicho is fabricated from one large flattened tin can cut to the configuration of the nicho. The large panes of glass forming the door and side panels are bound with narrow tubular strips of tin with quarter-round braces soldered at the corners. This device no doubt was intended to strengthen the structure of the nicho to prevent breakage of the glass when it was transported in processions. The nicho presently rests on a wooden litter (*anda*) that has been carefully constructed to hold this particular piece.

5.97

Soldered at the corners of the base are four loops of tin with ribbons drawn through them that are tied around the handles of the litter to secure the nicho when it is carried.

The nicho contains a bulto of the Virgin of Guadalupe which is attributed to the Santo Niño santero. While the bulto predates the nicho by as much as forty years, the nicho was undoubtedly made specifically for it. Flanking the bulto are two simple candle holders attached to the tin floor. Candle holders usually are designed to fit against the side panels on the outside of the body of a nicho, but in this unique design the candles were placed in the interior, more for ornament than utility. The heat and close proximity to the bulto would have either damaged the bulto or cracked the glass if the candles were burned for an extended time.

The Gonzales nicho is unusual, not only because of the dedication and signature but also for the conspicuous absence of worked tin ornamentation. Often, the characteristic designs on the tin have enabled us to recognize specific bodies of work; consequently, the limited use of decorated tinwork on this nicho is especially frustrating. The sole decorative tin element is a narrow scalloped band that defines the pediment and the adjacent upper sides, much like an architectural cornice. This narrow band has piqued our interest and curiosity. It is composed of a series of deeply cut, heavily embossed scallops along the upper edge that have been further decorated with a small circular stamp. This combination of stamping and embossing gives the

FIGURE 5.97
Detail of nicho, (left) signature roof panel, (right) dedication roof panel. H.V. Gonzales, dated 1872. Collection Archdiocese of Santa Fe.

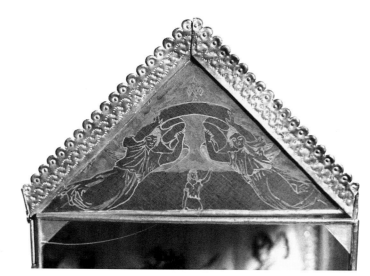

5.98

FIGURE 5.98
*Detail of nicho
pediment. H.V.
Gonzales, dated 1872.*

First, the deeply scalloped trim of the Gonzales nicho is nearly indistinguishable from the trim surrounding the lower portions of a nicho illustrated in E. Boyd's *Popular Arts of Spanish New Mexico* (page 299). The tin strip on the Boyd nicho is cut, stamped, and embossed in an identical style, and the same small circular stamp found on the signed nicho is used as a detail in the diaper pattern of the back panel. The Boyd nicho can be readily identified as an example of the work of the Rio Arriba Painted Workshop by its back-painted glass panes. The use of painted glass panels to create architectural form and the identical tin band suggest the possibility that H.V. Gonzales might have made the Rio Arriba Painted Workshop pieces. The vibrantly painted flowers on the panels of those pieces and the calligraphic roses of the Gonzales nicho indicate a relationship to a second example.

We must observe both the hand-painted flowered panels and the tinwork details on the complex box illustrated in Plate 4. The scalloped apron skirting the base of the box bears a striking similarity to the tin cornice of the Gonzales nicho. This tinwork detail in combination with the calligraphic drawing of flowers on the paper panels leads us to believe that Gonzales was very likely the maker of this box. The name lettered on the top panel of the box, "J. Atilano Suaso," is also quite similar to the varied script of the dedication and signature panels of Gonzales's nicho. In both the signed nicho and the trinket box we are aware of an unsophisticated but talented artist who was comfortable with paint and calligraphy and was able to utilize the two in intriguing combinations.

A third example is suggested by the lettering on

scallops a dimensional quality that resembles repoussé. The center of the band has been stamped with an undulating linear design created by two rows of offset stamping with a coarse, serrated crescent die. Because this decorated strip is one of the few distinctive details of the nicho, it must be regarded as a key to determining related pieces of tinwork.

Along with the decorative tin cornice, the other identifiable characteristics of the Gonzales nicho are the use of back-painted floral glass panels and the inclusion of lettering in the design of the panels. A comparison of these details with those of other New Mexican pieces suggests an interesting series of conjectures. Three examples that incorporate one or more of these characteristics into their design must be examined.

the top of the box. The frame in Figure 5.7 has been included in the section on the Santa Fe Federal Workshop, although there are stylistic details that are not typical of that workshop. It is the upright and pendant leaf forms that give this frame a Santa Fe Federal identity. As noted in the prior discussion of the frame in this chapter, the lower left appendage does not match the others. During the course of our survey we have seen several pieces of nineteenth-century tinwork whose parts have been "married." It is quite possible that this frame originally did not include the pendant and upright leaf forms. Some of the uncharacteristic details of the frame more closely resemble those of both the Gonzales piece and the Boyd nicho. The frame is elaborately decorated with a type of heavy stamping not typical of the Santa Fe Federal Workshop, which, along with the delicately scalloped corner fan pieces, suggests the type of stamping seen in both the tin cornice of the signed nicho and the work of the Rio Arriba Painted Workshop in general.

The most intriguing aspect of this frame is the hand-drawn and painted image of San Cayetano. Other New Mexican frames occasionally contain small drawings of the Santo Niño de Atocha executed in varying degrees of primitive style (see Figure 2.3). While this image of San Cayetano is executed in a somewhat naive manner, it is not primitive. The figure is convincing, and the whole picture is much more closely related to mid-nineteenth-century lithographs than are the small drawings of the Santo Niño. Finally, we must also examine the script at the bottom of the image in Figure 5.7. The words "S. Cayetano padre de la Providencia," similar to the lettering on the top panel of the box

and the roof panels of the Gonzales nicho, are executed with a sure, artistic hand, giving the impression of an artist in control not only of the medium but of the calligraphy as well. It is also important to know that this frame was for many years in the possession of the Indian potter, María Martínez, who lived at San Ildefonso Pueblo, adjacent to the Spanish village of the same name.

The design, calligraphic, and geographic relationships between these three examples and the large signed nicho are quite compelling. They suggest that H.V. Gonzales was an active tinsmith and an accomplished, creative craftsman who could have been responsible for much of the tinwork originating in the general vicinity of Española and surrounding Rio Arriba County during the 1870s. Until a more firm connection is established with a larger number of identifiable works, however, we must hesitate in assigning any of the other named workshops from northern New Mexico to him.

Other Tinsmiths

More than ninety percent of the nineteenth-century tinwork found in New Mexico can be attributed to the thirteen named workshops. A few extremely simple objects such as L-shaped sconces have little or no decoration that might provide clues to their identification. Additionally, works made primarily from glass and wallpaper panels (nichos, trinket boxes, or frames) often have so few tin parts that they cannot be linked to any recognized maker. Some pieces in this section were made by experienced craftsmen, though such a limited number of

5.99

5.100

FIGURE 5.99
*Frame containing
chromolithograph.
Unidentified
tinsmith, circa 1885,
20¹/₂ × 14³/₄.
Collection of Tony
Garcia.*

FIGURE 5.100
*Frame containing
four German Holy
Cards. Unidentified
tinsmith, circa 1890,
14¹/₄ × 10⁵/₈. Private
collection, Santa Fe.*

examples of their work exists that we are hesitant to assign them specifically. Others were obviously made by untrained craftsmen for their own use. A discussion of examples from both experienced (but unidentified) and amateur tinsmiths follows.

A well-made, elaborate frame by an unidentified tinsmith is illustrated in Figure 5.99. The frame contains a hand-colored European lithograph from the 1860s. The surface is extensively embossed with heavy single-dot punching in a design of rosettes and interlocking triangles enclosed by rows of blocks.

The lunette is decorated solely with single-dot embossing. The outer border of the lunette is cut into large scallops, while the corner rosettes have smaller scalloped edges that suggest composite flowers. The careful shaping of the parts and the measured stamping show that this piece was made by an experienced tinsmith. No other pieces of a similar nature have been located. This is one of the few articles of New Mexican tin with a well-documented collection history. The frame has descended in the family of the original owner,

5.101

5.102

Magdalena Gonzales of Bernalillo, New Mexico, and was presumably made there.

A rather crude piece by another maker is illustrated in Figure 5.100. This simple frame is butt-joined and lacks a tin backing plate. Four different German Holy Cards are mounted on a pasteboard backing. The edges of the side and bottom panels are serrated in the same manner as Taos Serrate work, but the unsophisticated single-dot-and-dash embossing is unrelated to surface decoration used by that workshop. The unusual lunette with two

protruding rabbit ears is embossed with a stylized pattern that resembles a dragonfly. This is one of four frames closely related in style and construction that are reported to have been collected in the Rio Abajo region. Although there are some stylistic similarities to major workshops, this maker was a naive craftsman who produced a limited number of frames for personal or local use.

By contrast, the frame containing a print of "San José" in Figure 5.101 was executed by an experienced hand. The side and end panels are butt-joined,

FIGURE 5.101
Frame containing hand-colored lithograph. Unidentified workshop, circa 1880, 22²/₃ × 14¹/₂. Collection of Will and Deborah Knappen, Santa Fe.

FIGURE 5.102
Frame containing Currier and Ives lithograph. Unidentified tinsmith, circa 1880, 17¹/₂ × 13. Collection of Hansen Gallery, Santa Fe.

5.103

FIGURE 5.103
Nicho. Unidentified
tinsmith, circa 1895,
26³/₄ × 17 × 4¹/₂.
Collection of Larry
and Alyce Frank.

are awkwardly proportioned and have minimal stamping. Containing a Currier and Ives lithograph of "The Holy Communion," circa 1875, this frame has been cut from a single large piece of tin with the center pierced to fit the print. The scalloped edges, pointed corners, and crude scoring that accents the interior opening are all reminiscent of characteristics of the Rio Arriba Workshop. The embossed stamping utilizes a unique die that makes the impression of multiple chevrons. A frame by the same maker (now in the Harwood Foundation Collection in Taos) is hand-stamped with a patent date of 1864. The dates of the prints that appear in this body of work suggest that it was produced circa 1870.

Numerous nichos exhibit construction methods and styles of ornamentation unrelated to those of any recognizable workshop. An example by an unidentified tinsmith is illustrated in Figure 5.103. The sawtooth borders of the side panels and lunette are extremely sharp, unlike the borders of Taos Serrate work. The well-executed minimal stamping has been produced with a limited variety of straight-bar stamps. The staccato effect produced by the sawtooth edge is repeated in the carefully scored rosette decorating the back panel of the nicho. Unusual spade-shaped finials that flank the lunette accentuate the geometric quality of the case section of the nicho.

Two large, ambitious, and complex nichos by an accomplished tinsmith are illustrated in Figures 5.104 and 5.105. The elongated form of both nichos, including the pediments and flanking towers, shows the influence of Victorian Gothic Revival architecture. A Rio Abajo origin is suggested by such details

a method that suggests early Rio Arriba work, although the unusual diagonally crimped border is not typical of that workshop. The carefully scalloped edge of the lunette is similar to the scalloped borders of both Mora Octagonal and Rio Abajo work. Scored lines radiating from the unusual spoked-wheel rosette also suggest the Rio Abajo style. It is possible that this frame represents an atypical example by one of the recognized workshops.

One style of frame that appears with some regularity is illustrated in Figure 5.102. Such frames

5.104

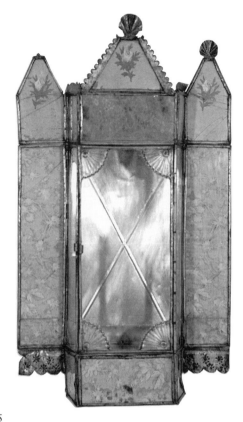

5.105

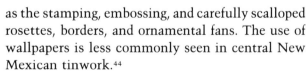

as the stamping, embossing, and carefully scalloped rosettes, borders, and ornamental fans. The use of wallpapers is less commonly seen in central New Mexican tinwork.[44]

The nicho in Figure 5.105 contains late-Victorian period wallpaper under glass in the side and bottom panels. The glass panel above the door features two layers of material, the wallpaper partially exposed under a swag of cloth. This combination produces the effect of a stage setting, as though a curtain were lifted to reveal the bulto. Elongated

pediments are capped with scored tin details resembling shells.

The treatment of finials in both nichos is unusually imaginative for New Mexican tinwork, and the extremely elaborate ones crowning the nicho in Figure 5.104 are unique. Two influences are represented in these finials. The floral design of the central finial is similar to tin ornaments that decorated elaborate late-Victorian residences. This suggests that the tinsmith may have had access to a carpenter's style manual or a catalog offering fac-

FIGURE 5.104
Nicho containing wallpaper and mirrors. Unidentified tinsmith, circa 1890, 41½ × 26 × 5½. Taylor Museum, Colorado Springs, Colorado.

FIGURE 5.105
Nicho containing wallpaper and cloth. Unidentified workshop, circa 1890, 26½ × 15¾ × 4. Private collection, Santa Fe.

5.106

5.107

FIGURE 5.106
Nicho containing hand-drawn Santo Niño de Atocha. Unidentified workshop, circa 1865, 13³/₄ × 7¹/₂ × 2. Collections of the Museum of International Folk Art, a unit of the Museum of New Mexico, Santa Fe.

FIGURE 5.107
Nicho. Unidentified tinsmith, Chamberino, Doña Ana County, New Mexico, circa 1870, 15 × 11¹/₂ × 9¹/₂. Collection of Mr. and Mrs. J. Paul Taylor, Mesilla.

tory-made architectural ornaments. The lower portion of the flanking finials shows an influence more directly related to urn shapes and floral arrangements in period prints, especially those of the Santo Niño de Atocha (a readily available prototype). It is also possible that the prototype for this nicho may have been an elaborate tabernacle or nicho of late eighteenth- or early nineteenth-century Mexican origin.

Both nichos are excellent examples of the toll that attrition and the passage of time have taken

on these fragile items. When they were photographed by Truman Mathews in the late 1930s, the glass in Figure 5.105 was intact and the cloth was darker and seemed to be more brilliantly colored. Today, it is badly damaged and the cloth panels have faded until the pattern is barely distinct. The nicho in Figure 5.104 has fared somewhat better. Nevertheless, part of the left finial is now missing and the mirrored rosettes (complete in the 1930s photograph) are cracked and broken.

The nicho in Figure 5.106 was originally in the

personal collection of the late E. Boyd.[45] The trapezoidal form, the engaged Solomonic columns, and the scalloped borders suggest a relationship to the Santa Fe Federal Workshop. The odd proportions may indicate that the piece originally had side panels. In this event, the proportions and the structure would be similar to those of other Santa Fe Federal nichos. The image of the Santo Niño de Atocha in the frame attached to the back panel is not a print but a watercolor drawing probably made as an assignment at the Loretto Academy in Santa Fe.[46]

The *hornacina* in Figure 5.107 is the only true example of an all-glass nicho. This piece is made entirely from glass panels edged in tin, with the exception of the roofs of the twin bell towers. Small brass bells still hang in each tower. The nicho is exceptional in its construction. It is completely three-dimensional, and the planning necessary to make the steeply pitched glass roof intersect with the hexagonal glass towers is astounding. Mounted on an altar surrounded by lighted candles, the nicho must surely have been an object of great devotion. The piece was made in Chamberino, Doña Ana County, in southern New Mexico about 1870. According to local tradition, the maker was known as Nuestra Señora de los Santos (Our Lady of the Saints). She is the only woman tinsmith reported in nineteenth-century New Mexico, but unfortunately over time her real name has been lost. There are no tinsmiths in the census for Doña Ana County during the period. The neoclassic-style Mexican frame with the small milagros attached to it was found in the nicho when it was acquired from Mrs. Luis Saucedo of Chamberino some twenty years ago.[47]

The small piece in Figure 5.108 is one example

5.108

of the many extremely simplified frames found in New Mexico with only a tin tube holding the glass pane in place over the Holy Card or print (in this example, a Mexican woodcut of St. Nicholas Tolantino). This frame was collected in the 1950s in the village of San José, San Miguel County.

The elaborate-style sconce shown in Figure 5.109 is one of a pair made for the Church of San Albino in Mesilla, New Mexico, which was built in 1854. The original church was replaced by a Romanesque brick structure in 1906. These sconces were later

FIGURE 5.108
Frame designed as a pendant containing a Mexican woodcut. Unidentified workshop, circa 1880, 4 1/8 × 2 5/8. Collected in San Miguel County, New Mexico. Collection of Mr. and Mrs. J. Paul Taylor, Mesilla.

5.109

FIGURE 5.109
Sconce, one of pair.
Unidentified
tinsmith, Doña Ana
County, circa 1860,
10⁵/₈ × 4¹/₂ × 5.
Collected in Mesilla,
circa 1880. Collection
of Mr. and Mrs. J.
Paul Taylor, Mesilla.

Certainly numerous pieces of New Mexican tin-work remain unrecognized and unidentified—pieces that do not fit comfortably into the styles of the named tinsmiths or workshops. Undoubtedly much of the unidentified work was made as single pieces by individuals who were unable to afford the work of the itinerant tinsmiths, or who lacked access to larger communities where they might obtain ready-made items. These skilled farmers, laborers, or even blacksmiths decided to fashion a tin frame, nicho, or sconce themselves. In such pieces the art of the tinsmith has often been simplified to its elemental state. The results range from the simple or primi-tive to the fanciful folk interpretations of the more complex and refined work that is regarded as clas-sic nineteenth-century New Mexican tinwork.

Although we have made every effort to weigh the reliability of the collection history of the pieces, it is possible that we were misled by works brought years ago from one part of the state to another by family members or by dealers. For example, a few pieces of tinwork made by the Valencia Red and Green II Workshop (in central New Mexico) have been found in Rio Arriba County and as far north as Taos. Similar problems concerning the Mesilla Combed Paint pieces and the Fred Harvey Com-pany buyers have already been noted in this chap-ter. Any mistakes involving provenance of the tinwork should be considered ours.

found in the former home of the resident priest. The form and stamping do not correspond to any of the work of known tinsmiths, but another sconce by the same maker was also collected in Mesilla and is now in the Gadsden Museum collection.

6
THE REVIVAL PERIOD

*B*y the beginning of the twentieth century, most of the workshops had ceased producing tinwork. Two exceptions are the Isleta Tinsmith, who continued to make large frames for chromolithographs and oleographs until the beginning of World War I, and José María Apodaca, who ceased working about 1915. Modernization during the early twentieth century brought immense changes to the old Spanish culture of New Mexico.

As the production of and demand for the nineteenth-century styles of tinwork declined, a series of events occurred that eventually led to a resurgence of tinwork—the Revival period. In the 1880s, California architects developed a new regional design form sympathetic to the Spanish missions built along the coast of California by Franciscan priests in the eighteenth and early nineteenth centuries.[1] This California Mission style of architecture became popular and was quickly accepted by the public. In New Mexico, the Fred Harvey Company built two railroad hotels in a modified version of this new Mission style, based on Spanish prototypes. The Alvarado was completed in Albuquerque in 1905 and the Los Chavez in Vaughn in 1909.[2] The El

Ortiz, built in Lamy in 1910, was the first hotel in the New Mexican Spanish Pueblo style.[3] A major development in the regional architecture of New Mexico began in 1905 when W.G. Tight, president of the University of New Mexico in Albuquerque, conceived a program to build a "pueblo-style University." By 1906, five buildings in the "pueblo-mission style" had been completed on the campus.[4] Several political figures and archaeologists founded the Museum of New Mexico in 1909 and set about restoring the Governor's Palace in Santa Fe to its "original" Spanish Pueblo form. The restoration of the palace from 1909 until its completion in 1913 was headed by archaeologist Jesse Nussbaum.

An additional impetus for the development of the Spanish Pueblo style of architecture was the New Mexico Building for the Panama-California Exposition in San Diego, designed by the Santa Fe architectural firm of Rapp and Rapp in 1915. The firm, originally from Trinidad, Colorado, had designed a large commercial building in Morley, Colorado, in the Spanish Pueblo style as early as 1908.[5] The design of the New Mexico Building was based on the Spanish Mission church and convent at Acoma Pueblo built in the second quarter of the seventeenth century. A year later, work began on the most imposing Spanish Pueblo Revival building in New Mexico: the Fine Arts Museum in Santa Fe, also designed by Rapp and Rapp, and built by Jesse Nussbaum.[6]

A number of artists from the East settled in Santa Fe before World War I. Among them were Carlos Vierra, Sheldon Parsons (the first director of the Fine Arts Museum), Gerald Cassidy, and William Penhallow Henderson. Following World War I, they were joined by Gustave Baumann, B.J.O. Nordfeldt, Willard Nash, Will Shuster, Randall Davey, and Frank Applegate. These artists, fascinated with the Spanish and Indian cultures of New Mexico, began to restore old homes or build new houses for themselves based on the Spanish Pueblo styles. Gerald Cassidy and Frank Applegate, in particular, furnished their homes with eighteenth- and nineteenth-century Spanish New Mexican furniture and nineteenth-century tinwork. In 1924, the writer Mary Austin and Applegate were instrumental in founding the Spanish-Colonial Arts Society, which was dedicated to the preservation of traditional Spanish arts.[7] The society was active until the early 1930s and introduced an annual Spanish Market in conjunction with the Santa Fe Fiesta in 1925 to encourage the continuation of traditional Hispanic crafts.[8]

Several Santa Fe artists were further involved in the promotion of indigenous crafts with the establishment of the Spanish and Indian Trading Company in 1925, managed by artist Andrew Dasburg.[9] The interest in both old and new Spanish crafts led to the establishment of the Spanish Arts Shop in 1930 by members of the Spanish-Colonial Arts Society, with primary support from Mary Austin. This was the first store exclusively devoted to the promotion and sale of new Hispanic Revival crafts. The shop closed in 1933, but was followed soon after by the most important shop of the period, the Native Market, founded by Miss Leonora Curtin.

The Native Market shop opened in Santa Fe during the Depression in June 1934 and expanded to include a branch store in Tucson, Arizona, in 1936. These stores provided outlets for Spanish crafts made by students of the county vocational schools, as well as other Hispanic New Mexican craftsmen. Goods sold included furniture, weaving, colcha stitching, ironwork, some religious carvings, and tinwork. Pedro Quintana, who had learned filigree jewelry-making from his father, Alejandro Quintana, was employed as the resident tinsmith.[10] The market also sold tinwork by Francisco Sandoval, Francisco Delgado, and later, Eddie Delgado and Frank Garcia, all of Santa Fe. It employed David Laemmle, an artist who painted glass panels for tin fixtures as well as decorating furniture. Mirror frames, candelabras, sconces, and lamps were the most popular tinwork sold in the shops. In spite of its popularity, the market unfortunately continued to operate at a loss and was always subsidized by Miss Curtin. In late 1939, a group of supporters organized the Native Market Guild and took over the ownership of the market, maintaining it as a crafts cooperative until 1940.[11] By this time the venture could no longer be subsidized, and the craftsmen had begun to find work related to the preparations for World War II.

Spanish craft stores brought about a shift from a traditionally Hispanic religious patronage to one in which buyers were often tourists, which greatly affected tinwork of the period. The new customers demanded pieces that not only had a "southwestern flavor" but also could be used in modern homes. Frames were made for mirrors instead of religious prints. Most wall sconces, chandeliers, and lanterns were electrified. New forms of tinwork were developed which had not existed in the nineteenth century. Table lamps, flowerpots, tissue holders, and ashtrays were new products of the Revival crafts-

men, made in response to the requests of twentieth-century buyers.

Another development in tinwork during the Revival period was the change from the use of salvaged tin cans to the almost exclusive use of another material—terneplate. Imported primarily from West Virginia, the new sheet material was used in place of new tinplate because of the mellow, darker color of the lead-tin alloy coating. In some ways, this dark tone imitated the oxidized finish that nineteenth-century tinwork had acquired. Additionally, its color was better suited for use in Spanish Pueblo Revival buildings and in restorations of older homes.

The first educational program devoted to Spanish crafts was started in Galisteo, New Mexico, by Concha Ortíz y Pino in 1929. The Colonial Hispanic Crafts School taught weaving, leatherworking, and furniture making.[12] El Rito Normal School began to offer some classes in traditional crafts as early as 1930. With the onset of the depression, a variety of New Deal programs allowed the development of an extensive county vocational school project, instituted by the state supervisor of vocational education, Brice Sewell. Sewell was appointed in 1932, and with the encouragement of Leonora Curtin modeled his schools on a grass-roots program developed by the Taos Crafts organization in 1933. This school became the Taos Vocational School, with the principal, Max Luna, teaching tinsmithing.[13] Luna was a descendant of a family of silversmiths and filigree jewelers from the Taos area.[14] In 1937 and 1938, the school produced the furnishings, including tinwork chandeliers, for the Harwood Foundation in Taos.

The county vocational program expanded quickly

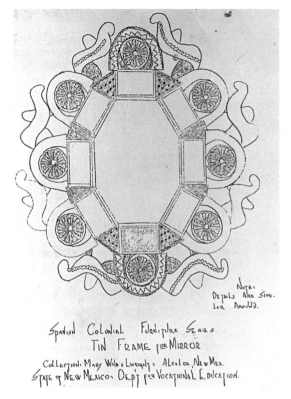

6.1

SPANISH COLONIAL FURNITURE SERIES
TIN FRAME FOR MIRROR
Collection: Mary Wheelwright = Alcalor, New Mex.
State of New Mexico = Dept for Vocational Education.

FIGURE 6.1
Drawing of tin frame from New Mexico Tin Craft, *1937, page 10. Private collection, Santa Fe.*

from the first school at Chupadero in Santa Fe County to more than twenty-five schools by 1936, extending from Socorro in central New Mexico to Costilla on the Colorado border north of Taos. The only schools that reported teaching tinsmithing, however, are the Taos program and possibly the small school at Puerto de Luna founded by Russell Vernon Hunter in 1934.[15] Hunter was later director of both the Federal Arts Project and the *Index of American Design* in New Mexico.

A member of Brice Sewell's staff, Carmen Es-

pinosa, wrote *New Mexico Tin Craft*, one of the "blue books" published by the Vocational Department to accompany the volumes on furniture, colcha embroidery, hide tanning, and others. This mimeographed book, published in 1937, included drawings by Truman Mathews of more than twenty-five nineteenth-century tin pieces, along with an introductory text. The book focused primarily on drawings of frames (Figure 6.1), but included a chandelier, three electrified sconces, two nichos, and a cross. It featured work from nine of the nineteenth-century workshops, excluding the Small Scallop, Taos Serrate, and Santa Fe Federal styles. Though the book was apparently little used by the vocational schools, local tinsmiths utilized some of the illustrations to develop their own styles.

By 1940, the mood and focus of the country had changed. Both availability and interest in the Spanish crafts waned as the depression ended and the country prepared for war. The crafts people trained by the vocational schools left for jobs in war plants, particularly on the West Coast, or began to volunteer for military duty. The programs had accomplished their two most important goals, however; to help the people of New Mexico survive the depression and to revive the old Spanish craft forms of New Mexico.

The products of the Revival period were distinctly different from their nineteenth-century predecessors in both form and function. During the Revival the individual tinsmiths often randomly adapted design elements from the nineteenth-century workshops and from Mexican tinwork into an eclectic southwestern style rather than using a style unique to a particular tinsmith. This eclecticism

6.2

may have been due in part to increased awareness and exposure to the early collections formed by Frank Applegate, Mary Cabot Wheelwright, Leonora Curtin, Gerald and Ina Sizer Cassidy, Pamela Parsons, and others with a broad interest in the Spanish Southwest. These first collectors assembled a broad range of objects from furniture to farm tools, as well as native religious objects including bultos, retablos, and tin nichos and frames. In addition to private collections, a number of shops established in the late 1920s also displayed nineteenth-

century tinwork. These collections provided a wealth of material for the Revival craftsman.

The isolation of the nineteenth century that had contributed to the individualized designs of the early workshops was no longer a factor, and a new style of tinwork developed that continues into the present. The Revival period saw the design vocabulary of nineteenth-century tinwork adapted to twentieth-century requirements. Besides the popular Santa Fe Federal style, the workshops most often used as a design source by Revival tinsmiths were

FIGURE 6.2
Francisco Sandoval's tinshop (ojalateria) on Water Street, Santa Fe, 1916. Unidentified photographer. Photo courtesy Museum of New Mexico, negative number 23125.

6.3

6.4

FIGURE 6.3
*Cross-sconce.
Francisco Sandoval,
circa 1930, 12 × 10,
terneplate, glass, and
wallpaper. Collection
of Dwellings
Revisited, Taos.*

FIGURE 6.4
*Frame with mirror.
Francisco Sandoval,
signed on reverse
"F.W. Sandoval," circa
1930, 16 × 14,
terneplate. Private
collection, Santa Fe.*

the Valencia Red and Green, Valencia Red and Green II, Rio Arriba Painted, and Mora Octagonal. A discussion of the work of the most notable Revival tinsmiths will show that while many nineteenth-century motifs were utilized, the major tinsmiths did develop work of individual character nevertheless.

Francisco Sandoval of Santa Fe (1860–1944) was a craftsman whose career spanned the period from the late nineteenth century into the first half of the twentieth century. Sandoval was a tinner, owned a plumbing shop, and (according to Espinosa) remembered buying lard cans from the army as a young boy (Figure 6.2).[16] Several signed or readily identifiable pieces by him from the 1920s and 1930s period still exist, but the work he produced in the late nineteenth century remains unidentified. Sandoval supplied tinwork to the Spanish Arts Shop and later to the Native Market. Stylistically, his later work bears a resemblance to the work of the Rio Arriba Workshop. It could be that he, and perhaps his father (also a tinsmith), are the makers of that extensive body of work.[17]

Two distinct examples of the work of Francisco Sandoval are the cross-sconce and framed mirror in Figures 6.3 and 6.4. Both examples are constructed of terneplate and represent the excellent craftsmanship and design characteristic of Sandoval's work. Though the cross is based on a nineteenth-century form, it is composed of design elements typical of his style. The use of trefoil scallops, the repeated concentric circles of the finial, the side rondels, and the shape of the base are commonly seen in Sandoval's work. The use of applied convex discs is also a typical detail. Another decorative device used by Sandoval is continuous bands of heavily stamped V's that appear at the top of the cross below the trefoil finial and at the base of the cross. The frame includes many of these details. A similar crest (not a lunette) with the same trefoil design is embossed with a rosette flanked by two convex appliqued discs, and the shape of the side scrolls is typical of many Sandoval pieces. The extensive deep embossing and stamping gives this frame its robust character.

The candelabra in Figure 6.5 is representative of many produced in the 1920s and 1930s by Revival tinsmiths. This candelabra was identified by Pedro Quintana as a design particular to Francisco Sandoval in this period.[18] It is notable that this form of candelabra, not seen in nineteenth-century work, was one of the Revival designs that seems to have evolved in response to patrons' demands for objects that were functional as well as decorative.

Francisco Delgado (1858–1936) must have been an interesting individual. He learned tinsmithing around the turn of the century and produced tin-

6.5

work for more than twenty years on a part-time basis. Listed as a Santa Fe merchant in the 1910 census, he later became a stenographer and book-keeper for several federal and state agencies. After his retirement from government work in 1929, he became the proprietor of the Colonial Tin Antiques shop on Canyon Road in Santa Fe.[19] A photograph, circa 1935 (Figure 6.6), shows Francisco Delgado at work in his Canyon Road shop, surrounded by his own tinwork and his collection of nineteenth-century examples. The popularity of his work is ap-

FIGURE 6.5
Candelabra.
Francisco Sandoval,
circa 1935, 10³/₄ ×
17, base 6-inch
diameter, terneplate.
Private collection,
Santa Fe.

6.6

parent from the many pieces in various stages of completion.

Figure 6.7 shows a lovely example of Delgado's work. Two recognizable nineteenth-century design influences are apparent. The first is the use of Federal-inspired engaged columns to form the body of the frame. The second is the use of two side pendants that are derived directly from the Rio Arriba Workshop tradition. The crest is ornamented with a handsome rosette and excellent stamping.

The framed mirror in Figure 6.8 is a variation, often repeated, of a Valencia Red and Green II frame (see Figure 5.77). A nearly identical frame appears in the photograph of Delgado's studio. Valencia Red and Green II work appears to have been a favorite resource for Delgado, and several pieces based on these designs still exist.

Although not as well recognized as his father, Ildeberto "Eddie" Delgado (1883–1966) was an outstanding tinsmith for more than forty years. Like his father, he possessed a broad range of talents and interests. He is listed variously as the proprietor of

6.7

6.8

Delgado's Curio Shop in Santa Fe in 1928, as a tin-smith producing work for El Parian (the Native Market) in 1937, and as a tinsmith in the 1940s. In the 1950s he designed and constructed wall sconces for the restoration of San Miguel Chapel in Santa Fe.[20] Two examples of work by Eddie Delgado are the large standing cross in Figure 6.9 and the nicho in Figure 6.10. The cross, although extremely simple, is well made, with a monumental power and dignity not always seen in Revival work. The design is influenced by the Valencia Red and Green

and Rio Arriba workshops. The base is embossed with a simple rosette, similar to the cross in Figure 4.22, and the border is stamped with a tight chain of crescent stamps in the manner of the best Valencia Red and Green Tinsmith pieces.

The delightful electrified nicho in Figure 6.10 is an excellent example of the eclectic nature of Revival tinwork. Representing one of several lighting fixtures created by Eddie Delgado for the Albuquerque Community Playhouse, this example shows three distinct stylistic influences. The body

FIGURE 6.7
Frame containing mirror. Francisco Delgado, circa 1925, 11 × 13¹/₂, terneplate. Collection of Mr. and Mrs. John Conner Hill, Phoenix, Arizona.

FIGURE 6.8
Frame with mirror. Attributed to Francisco Delgado, circa 1935, 27 × 20¹/₂, terneplate. Collection of Dwellings Revisited, Taos.

6.9

6.10

FIGURE 6.9
*Cross on base. Eddie
Delgado, metal tag
on reverse: "E.
Delgado," circa 1935,
28 × 16¹/₂,
terneplate. Cross: 64
× 14, Base: 7¹/₂ ×
5¹/₄ × 5¹/₄. Private
collection, Santa Fe.*

FIGURE 6.10
*Wall sconce by Eddie
Delgado for The
Albuquerque
Community
Playhouse,
Albuquerque, circa
1935. WPA
photograph by
Russell Vernon
Hunter, courtesy New
Mexico State Records
Center and Archives,
Sante Fe, R.V. Hunter
collection #23016.*

of the nicho is similar in form and decoration to those of the Rio Arriba Painted Workshop, the side pieces are closely related to examples from the Valencia Red and Green Tinsmith, and the elaborate crest perched on top of the nicho is complete with speckled birds characteristic of the Valencia Red and Green II Workshop. Certainly Delgado, as proprietor of a shop selling curios and antiques, must have had a wealth of tinwork from which to choose the design motifs that were combined in this fanciful creation.

Benjamin Sweringen (active 1928–36) was a man of varied interests, like his contemporaries the Delgados. Although he trained as a painter and exhibited a painting in the Fiesta exhibition at the Fine Arts Museum in Santa Fe in 1924, it is not known how he learned tinsmithing.[21] With the opening in 1928 of his Santa Fe shop, The Spanish Chest, he provided a marketplace for furniture, lighting fixtures, and other household items with a southwestern flavor. The work of this talented craftsman is often confused with that of his younger contem-

6.11

6.12

porary, Robert Woodman. Their work, produced in the late 1920s and the 1930s, is very similar in both style and execution. A page from *School Arts* magazine, dated 1934, shows three chandeliers and a mirror with attached light sconces.[22] Attributed to Mr. "Sweringer" [sic], these examples illustrate his ability to fabricate both simple and complicated tinwork. One of the chandeliers hangs in the foyer of the Laboratory of Anthropology in Santa Fe, which was built 1929–33. The large piece is constructed of multiple parts: extensively stamped tin panels,

pierced tin, and painted and combed glass panels. Both this chandelier and those hanging in the Santa Fe County Court House represent the most elaborate known examples of Sweringen's work. Less elaborate but equally charming is his small framed mirror seen in Figure 6.11. Made about 1930, the design of this terneplate frame was copied directly from a nineteenth-century example by the Rio Arriba Workshop.

Pedro Quintana, born in 1910, was trained as a filigree jeweler by his father, Alejandro Quintana,

FIGURE 6.11
Frame with mirror. Benjamin Sweringen, circa 1935, 10¾ × 9, terneplate. Collection of Dwellings Revisited, Taos.

FIGURE 6.12
Frame containing mirror. Pedro Quintana, circa 1935, terneplate. Collection of Mr. and Mrs. Pedro Quintana, Santa Fe.

6.13

6.14

FIGURE 6.13
*Electrified nicho.
Pedro Quintana and
David Laemmle,
circa 1935, 10¹/₂ × 12
× 4. Private
collection, Santa Fe.*

FIGURE 6.14
*Sconce with mirror.
Robert Woodman,
circa 1940, 18 × 18
× 3¹/₂, terneplate.
Private collection,
Santa Fe.*

and began tinsmithing in 1933 at the request of Brice Sewell, director of vocational education for the State of New Mexico. Mr. Quintana was the resident tinsmith for the Native Market from 1934 to 1938.[23] The large framed mirror in Figure 6.12 represents Quintana's first effort as a tinsmith.[24] His training as a jeweler and craftsman is apparent in this example. Beautiful in execution as well as design, the frame exhibits a strong influence of the Federal style, with neoclassic details and a lunette that more closely resembles Mexican tinwork.[25] The

skillful combination of scoring, stamping, and embossing shows the degree of care and craftsmanship with which Mr. Quintana approached his craft.

Figure 6.13 illustrates an electrified nicho by Quintana, with reverse-painted and combed glass by another Native Market craftsman, David Laemmle.[26] The form and design of the nicho is directly related to those produced by the Rio Arriba Painted Workshop (Plate 6). In Quintana's nicho the stamping has become extremely simplified, with an emphasis on the solid, well-constructed form of

6.15

the piece. The nicho was electrified originally and was designed to be hung as a movable light fixture. The original label, reading "Spanish Artisans of New Mexico," remains on the reverse of the nicho. The reverse-glass painting by Mr. Laemmle is executed in the manner of the Rio Arriba Painted Workshop, although the style of combing and the color selection of orange, black, blue, and white are more closely related to the artistic preferences of the 1930s than to the colors in the nineteenth-century prototype.

The career of Robert Woodman (1908–83)

spanned five decades, commencing in the late 1920s. He often worked with the noted Santa Fe architect, John Gaw Meem. Today, their combined efforts may be seen in public buildings such as the Laboratory of Anthropology (built in 1930) and the Zimmerman Library at the University of New Mexico (built in 1936). Other major work by Woodman can be seen at Bishop's Lodge near Tesuque, New Mexico, in Our Lady of Carmel Church in Montecito, California, and in numerous residences throughout the Southwest. Woodman's work is extremely durable

FIGURE 6.15
Photograph of Robert Woodman soldering helmets. Photographer unknown, circa 1940. Collection of d'Anne Woodman, Albuquerque.

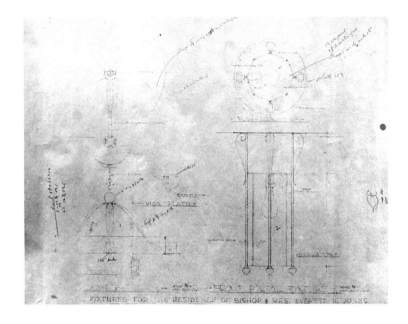

6.16

FIGURE 6.16
*Detail of drawing by
the firm of Meem and
Zehner for lighting
fixture, 1951. Private
collection, Santa Fe.*

and well-crafted, due in part to his training as an engineer. He drew inspiration from many sources. Especially evident in his work of the 1930s and 1940s is an extensive Valencia Red and Green vocabulary that appears in mirrors, wall sconces, and chandeliers. Woodman's mature work (after 1940) included more original designs, incorporating such varied materials as marbles, colored mirrors, and the extensive use of painted glass.

The electrified wall sconce with mirror in Figure 6.14 is an example of Woodman's ability to utilize nineteenth-century ornamental elements while creating a work uniquely his own. The multiple small rosettes and larger corner rosettes, typical of the Valencia Red and Green Tinsmith, are incorporated in much the same manner as the original. The central disc is scored as a radiating sunburst encompassing a mirror to reflect the electric light in the way candlelight was reflected in a nineteenth-century piece. The simple stamping is carefully executed. Closely placed inverted crescents are the predominant surface ornamentation.

Figure 6.15 offers a rare glimpse into an active Revival tinshop, circa 1940. It shows the broad variety of work produced in the Revival period. In this photograph, Robert Woodman is fashioning tin helmets to be worn by Los Caballeros de Don Diego de Vargas during the annual Santa Fe Fiesta.

The relationship between Revival craftsman and architect is most evident in works by Woodman. Figure 6.16 shows a construction drawing by the firm of Meem and Zehner for a residential entry fixture designed in 1951. Woodman's construction notes on this drawing enabled him to transform the flat drawing into a three-dimensional tin fixture. The chandelier in Figure 6.17 shows Robert Woodman's work at its best. This elegant fixture, composed of graceful curves and carefully balanced proportions, hung for many years in the Woodman home in Santa Fe. Supporting structural members are decorated with modified tin acanthus leaves, carefully cut, scored, and attached to the frame of the chandelier. Although not visible, the interior of the band that supports the lights is also decorated. Electrified components have been carefully concealed within the structural parts, or with orna-

6.17

6.18

mental tin elements. These are integrated cleverly into the design, so that we are aware only of the beauty rather than the functional nature of the piece.

Bruce Cooper (1905–87) was a late participant in the Revival. Beginning in 1936 he worked with Sweringen at the Spanish Chest, and he was involved later in furniture and lighting design and manufacture. He was a collector as well as a dealer in nineteenth-century tinwork and other New Mexican folk crafts. As a tinsmith, he had a flair for the theatrical and flamboyant. His colorfully painted

glass and tin creations were popular in the late 1940s and 1950s and continue to be collected today.

The large framed mirror in Figure 6.18 is one of the finest examples of Cooper's work. Composed of numerous tin and lead pieces, the floral elements under glass panels are carefully placed against a crushed blue velvet background. This frame is more closely tied to Spanish baroque design concepts than is the work of any other twentieth-century tinsmith, and it is much more elaborate and florid than examples from the nineteenth century.

FIGURE 6.17
Chandelier. Robert Woodman, circa 1940, 18 × 24 diameter, terneplate. Private collection, Santa Fe.

FIGURE 6.18
Frame with mirror, terneplate, velvet, and lead. Bruce Cooper, circa 1945, 45¹/₄ × 28¹/₂. Collection of Tukey Cleveland, Santa Fe.

6.19

FIGURE 6.19
Frame with painted glass, "Corazon de María." Pedro Cervantes, circa 1935. Photo courtesy R.V. Hunter Collection #23065, New Mexico State Records Center and Archives, Santa Fe. Frame now in the collection of the Museum of International Folk Art, a unit of the Museum of New Mexico, Santa Fe.

tinsmith is the tin frame with a painted glass panel of "Corazon de María" seen in Figure 6.19. The only known work of Pedro Cervantes, it was made under commission from the Federal Arts Project. It is based on designs of the Taos Serrate Tinsmith, with the corner pieces executed in the same manner as those in Figure 5.22 and the side panels stamped with the characteristic chevron pattern of the Taos Tinsmith (see nicho, Figure 5.21). It is probable that Cervantes produced this example as a copy of a nineteenth-century Taos Serrate frame. The reverse-painting on glass is well executed in brilliant colors against a creamy white combed-paint background.

There were other craftsmen producing tinwork during the 1920s and 1930s who have been forgotten or who worked with tin for only a brief period. No doubt other tinsmiths' work will appear and they will receive deserved credit as further research is done in this area.

The Revival period was a phenomenon that is unlikely to be repeated, though some tinwork based on both the nineteenth-century and the Revival style is being continued today by descendants of the earlier tinsmiths. The major craftsmen of the period were so prolific and so popular that their work has always been recognized as the standard for Revival tinworking in New Mexico.

Juan Garcia, a tinsmith from San Miguel, won a prize in the Third Annual Spanish Market in 1927 for a tin nicho.[27] Nothing else is known of his work. One interesting example by another little-known

7
IDENTIFYING MEXICAN TINWORK

During our search we have located over five hundred pieces of Mexican-made tinwork, primarily frames and nichos. These have been discovered in both public and private collections, although the majority are privately held and have been imported recently across the border at Juarez. Old collections of tinwork in New Mexico include few Mexican pieces, leading us to believe that most Mexican tin in this country has been collected in the last twenty years. Since the bulk of the work was purchased from dealers in the border towns, it is difficult to trace their origins or to determine if these pieces are fully representative of Mexican tinwork production. Interviews with dealers who have collected from homes in Mexico have been helpful in determining at least some areas of production of nineteenth-century tinwork in Mexico.[1]

While there has been considerable focus on the painted metal images called retablos, very little attention has been paid to the tin frames that embellish them. At first the similarities between New Mexican and Mexican tinwork make it difficult to distinguish between the two. There are distinct and recognizable differences which provide the basis for separate identification, however. It is important first to review briefly the development of tinwork in Spain and Mexico.

Tinworking developed at a rather early date in Spain, since tin plate from Germany was available to the Spanish in the seventeenth century. British tinplate was imported as early as 1750, and by the late eighteenth century Spain produced its own sheet tin in limited quantity.[2] Boxes of tinplate were compact and easily shipped as ballast in ships destined for Mexico and other parts of the Spanish empire. Thus, it is possible that the craft of tinsmithing developed in Mexico during the eighteenth century.

As we mentioned in Chapter 1, an early reference to tin objects appears in the Domínguez inventories of missions in New Mexico in 1776.[3] At least five processional crosses were found, along with host boxes and tin boxes for holy oils. Six tinplate candlesticks were reported from the church at San Felipe Pueblo. Domínguez also mentioned one tinplate lighting fixture at San Ildefonso: "A very pretty tin-plate lamp hangs from the vaulted arch. It is of medium size and is drawn up by a silk cord. . . . Father Tagle provided everything mentioned from the tabernacle to the lamp."[4] These pieces almost certainly originated in Mexico or Spain and were subsequently transported to the mission churches under the auspices of the clergy or the government.

Devotional images painted on copper sheets became popular in Mexico during the late eighteenth century, but by the 1830s tin plate began to replace the copper sheets.[5] One early example of Mexican votive (ex-voto) painting on tin is dated 1825.[6] Both retablos and ex-votos were painted on new tin plates imported from England and are found in standard sizes, primarily ten by fourteen inches. Less than 10 percent of the tin retablos were originally enclosed in ornate frames or shallow nichos, as is indicated by the number of tin retablos with remnants of solder on the back, although some frames were made with an opening at the top that allowed the retablo to be slipped into place.[7] Many were either framed in wood or had no frame, judging by the nail holes made when attaching the painting to a wall. Occasionally, tin frames are found with the retablo soldered in place; rarely, the image is found painted directly on the inside back panel of shallow nichos. The majority of the tin frames, however, like their New Mexican counterparts, were made to house and protect devotional woodcuts, engravings, and lithographs.

There are a number of subtle differences between nineteenth-century Mexican and New Mexican tinwork. The differences are not absolute but generally should be considered the keys to identifying the two. Usually absent from nineteenth-century Mexican tinwork are embossed can labels, seams, or printed labels, indicating that most of the pieces were made from new tinplate, not from salvaged containers. For the most part, the nineteenth-century Mexican tinwork found today either has oxidized to a dark grey or is extensively rusted. Tinplate used to fabricate nineteenth-century Mexican tinwork was the thinnest grade available (IC, or "one common") and has the least amount of tin coating of any of the grades of tinplate, thus allowing the natural oxidation of the tin surface to begin to expose the dark iron beneath. Even well-preserved tinwork frames with both original glass and paper prints intact are often rusted or heavily oxidized on their exposed surfaces.

One form of tinwork is peculiar to Mexico—the shallow nicho or deep frame. These pieces are made with the same designs used in flat frames, ornamented with side panels and a pediment or lunette on the top, but with a shallow tin box fitted to the back. The box creates a space about one inch deep between the glass and the back panel. Commonly, the glass pane was made removable by leaving a narrow slot at the top of the box or by hinging the front panel. Access to the interior of the shallow nicho facilitated the periodic changing and additions of devotional images, paper or cloth flowers, and small silvered milagros. With the exception of the work of the single border craftsman, the Mesilla Combed Paint Tinsmith, the shallow nicho-frame is rarely seen in New Mexican tinwork.

Mexican tinsmiths were generally more accomplished craftsmen than their New Mexican counterparts, and their range of decorative techniques was more extensive. True repoussé, or three-dimensional hammering, is common in Mexican tinwork, as is the difficult technique of curved scoring accomplished with a variety of templates. Templates were also used for reproducing side scrolls and pendant leaf shapes. Repoussé and scoring were used for floral and organic designs, particularly on lunettes and pediments (Figure 7.1). We have examined numerous examples of repeated design motifs and duplicate parts, which suggests that much Mexican tinwork was produced in organized regional workshops in the same manner as it was in nineteenth-century New Mexico.

Often the techniques of repoussé and scoring

7.1

were combined with piercing to create openwork areas that were further embellished with stamping and embossing. The Mexican tinsmiths used a greater variety of stamping dies than the tinsmiths of New Mexico and consequently developed a broader vocabulary of complex surface patterns. However, the side appendages, scrolls, and leaf forms found on Mexican tinwork are generally more simplified, abstract, and geometric than those seen in New Mexican tinwork. One small construction detail unique to Mexican tinwork that may serve as a key to identification involves the hanging loop

FIGURE 7.1
Frame containing painted tin retablo. Mexico, circa 1875, 16³⁄₈ × 12³⁄₄. New Mexico State University Art Gallery, Las Cruces.

7.2

duced primarily from Spain in the eighteenth and nineteenth centuries. The style that had the most effect on the nineteenth-century tinsmiths was the neoclassic, which began about 1785 and continued into the mid-nineteenth century.

The similarities between Mexican and New Mexican tinwork frames are striking, probably owing to a variety of parallel design sources and nearly identical metal-working techniques. Frames undoubtedly evolved in nineteenth-century Mexico, as they did in New Mexico, in response to the need to protect and embellish the inexpensive devotional prints that were available to the general population. The size of the devotional print or instructional Holy Card usually determined the size of frame. The popularity of colored lithographs eventually led to a decline in the production of retablos in Mexico.

Mexican frames are constructed of narrow ornamented tin strips or glass panels joined at the corners by a variety of fan shapes or square corner bosses. With few exceptions, some form of crest—a semicircular lunette or triangular pediment—is found at the top of the frames.

Large nichos, candelabras, and chandeliers are forms of nineteenth-century Mexican tinwork rarely seen in American collections, and few trinket boxes are known. As a result, it is difficult to make general observations about the similarity between these objects and their New Mexican counterparts. The large Mexican nichos known to us generally use the basic trapezoid shape and are fabricated of expansive glass panes that allow maximum exposure of the object of veneration.

The following discussion of selected works of

soldered to the back panel. In Mexican examples, the sharp edges of the tin support loop are carefully folded over, providing greater strength, while in New Mexico the hanging loop usually remains a single thickness of tin. In summary, then, distinguishing characteristics of Mexican tinwork include a dark coloration of the surface, the use of a deep frame, extensive repoussé and piercing techniques, more complex floral images, and the general absence of salvaged tin.

Mexican tinwork designs of the nineteenth century reflect a variety of architectural styles intro-

known Mexican origin should be helpful in determining some of these stylistic differences. Two outstanding examples of early nineteenth-century Mexican frames are shown in Figures 7.2 and 7.3. The simple yet elegant frame in Figure 7.2 is designed with neoclassic elements. Rectangular in form with no evidence of a lunette, the frame is composed of diagonally scored engaged columns placed against a stamped tin background. The columns are joined at the corners by four square corner bosses, each with a minimum of ornamentation. The prototypes for this type of frame were undoubtedly the restrained neoclassic wooden mirror and picture frames of the late eighteenth and early nineteenth century. The frame contains a mid-nineteenth-century portrait on tin by an anonymous provincial academic painter working in the style of José María Estrada.

Similar in concept but more elaborate in execution is the frame containing a Mexican engraving of "San Ysidro Labrodor," dated 1817 (Figure 7.3). The basic form, again, consists of scored, engaged columns joined at the corners by square corner bosses. The severity of the frame is softened by the oversize lunette and attendant side appendages. Rosettes ornamenting the corner bosses are repeated as one large rosette in the center of the lunette. Scored lines suggesting a sunburst radiate from the central rosette. This frame shows no evidence of repair or restoration, and both the glass and the print appear original. Although the length of time the engraving was in circulation is not known, we believe the frame was fabricated between 1817 and 1840. A nearly identical frame with a print of Nuestra Señora de Guadalupe is in the collection of the east morada of the Penitente brotherhood in Abi-

7.3

quiú, New Mexico.[8] It represents one of the very few Mexican examples found in old New Mexican collections.

A frame containing a devotional print of similar age to that of the previous frame is illustrated in Figure 7.4. The scored fan lunette and side appendages are similar, but the body of the frame is formed by glass panels joined at the corners by scored, fan-shaped bosses. The application of narrow strips of gold foil over tissue as well as the gold foil surrounding the figure of "Santa María de Atocha" provide an elegant touch to an otherwise simple frame.

FIGURE 7.3
Frame containing Mexican lithograph. Mexico, circa 1820, 17³/₄ × 14¹/₂. Courtesy Galeria San Ysidro, El Paso, Texas.

7.4

7.5

FIGURE 7.4
*Frame containing
Mexican lithograph.
Mexico, circa 1830,
18 × 14. Private
collection, Santa Fe.*

FIGURE 7.5
*Frame containing
confraternity
certificate. Mexico,
circa 1895, 13 × 9 1/4
× 3/4. Collection of
H. Malcom Grimmer,
Santa Fe.*

Two frames probably dating from 1860 to 1880 are those seen in Figure 7.1 and Plate 15. The first is an excellent example of fine and extensive use of curvilinear scoring, repoussé, and piercing. The organic nature of the design in the lunette and the rendering of the leaf and flower forms are typical of this type of Mexican craftsmanship. The sophistication of design and excellent filigree work in this example have not been seen in any New Mexican tinwork. The frame in the second example is similar in execution and construction, but is noteworthy for the realistic interpretation of the floral motifs that are reverse painted on the glass panels. This type of naturalistic painting is commonly seen in Mexican reverse glass painting, whereas floral motifs in New Mexican painted glass generally are more abstract.

Another frame that derives its interest from colorful reverse-painted glass panels is illustrated in Figure 7.5. This small, shallow nicho-frame contains an original Confraternity Certificate dated 1895. We have located as many as twenty-five frames

7.6

7.7

with painted glass panels nearly identical to this one. The specific choice of colors and a somewhat idiosyncratic method of combing the paint suggest that these nicho-frames were the product of a single workshop.

The charming frame in Figure 7.6 represents a very common stylistic type. As many as one hundred similar frames have been found, each bearing repeated motifs and/or identical proportions and measurements. This, again, suggests the probability of a specific workshop with numerous employees.

Angular and geometric in concept and form, these shallow nicho-frames are relatively restrained, perhaps showing some influence of the Gothic Revival style of the mid-nineteenth century. In this instance, the flaring side fins seen in the frame in Figure 7.3 have been reduced in this example to diminutive appendages and the expressionistic fan lunette has been replaced by a triangular pediment. The pediment is stamped and embossed with a floral motif composed of one centrally placed composite flower flanked by two naturalistic leaves

FIGURE 7.6
Frame containing daguerreotype. Mexico, circa 1860, 13 1/2 × 8 1/4 × 3/4. Private collection, Santa Fe.

FIGURE 7.7
Frame containing Holy Card. Mexico, circa 1885, 8 × 5 x 1. Collection of Mr. and Mrs. Murdoch Finlayson, Santa Fe.

produced by scoring. The severe and restrained nature of this frame focuses the attention on the image within it, in this example a daguerreotype from San Luis Potosí showing two statues, one in a flowing gown covered with milagros. The style of dress and the age of the photograph suggest a date circa 1860–70. Usually, frames of this type contain late nineteenth-century oleographs from the Benziger firm or other colored prints made in the last quarter of the nineteenth century. In this example, the daguerreotype appears to be original to the frame, although it is not possible to determine whether the frame was made specifically for the photograph. We must assume that either the photograph predates the frame or the workshop flourished for an extended period of time.

An example of the most common shallow nicho-frame is illustrated in Figure 7.7. Ornamented with a minimum of stamping, these simple frames were made for Holy Cards or other small devotional items. The lunette is scored and decorated with single-dot embossing, and the side scrolls are simple and abstracted.

The three-dimensional nicho illustrated in Plate 16 is a magnificent example of sophisticated nineteenth-century Mexican tinsmithing. Actually an *ofreta*, or offering box, the entire piece has a distinctly architectural character reminiscent of the many domed colonial churches of Mexico. Collected in Puebla, this superb example of Mexican tinwork is composed of many architectural features from the carefully executed dome to the well-crafted dentil cornice moldings. The dome is surmounted with a draped cross cut from tin. Four curious spool-shaped finials on the cornice suggest candle holders, though they are closed at the top and probably were made solely as ornament. A glass-enclosed pediment in front of the dome contains a rich assortment of gilt paper and silk cloth. The entire piece was originally painted and retains vestiges of red, green, and yellow pigment on the exterior, while the interior is painted a rich blue that provides a lovely background for the bulto. This example shows architectural form translated into tin by a skilled craftsman. While architecture was also a source of inspiration for nichos made by New Mexican tinsmiths, their work was always more simplified in form and stylized in ornamentation.

Although nineteenth-century New Mexican and Mexican tinwork can be differentiated in terms of style and degree of sophistication in execution, it is apparent that there were common sources of inspiration that reached beyond the necessity to protect and decorate devotional images. The two cultures shared European design styles dating from the seventeenth century, a long history of accomplished metalworking, a love of surface embellishment, and the importance of devotion in the Spanish Catholic culture.

While its production was independent, Mexican tinwork of the early nineteenth century must have provided the initial impetus for the tinwork subsequently produced in New Mexico, given the trade and cultural ties between them. Perhaps a few frames or nichos came north via the Chihuahua Trail to provide models for the craftsmen of "la frontera." Future research on Mexican and Spanish tinwork may provide clues to the connections and prototypes that created similarities between the two distinct groups.

CONCLUSION

We have surveyed the forms of tinwork made in New Mexico and noted the relative scarcity of various types. We have identified thirteen major workshops and described their design characteristics. In establishing the general locations for these workshops, we have been able to develop a series of names for them. We have reviewed the work of the Revival period tinsmiths and proffered some keys to identifying Mexican tinwork.

Other areas still need further study for a better understanding of the design development of tinwork in New Mexico. A thorough search of the collections of tinwork in Mexico would yield information about the inception and changes in tinsmithing in that country, and dated pieces from the first half of the nineteenth century would show the prototypes that were available to the first craftsmen in New Mexico. A study of the eighteenth- and nineteenth-century tinwork of Spain and possibly North Africa would also add to our knowledge of the growth of tinworking in the Spanish world. A more complete understanding of the connections between the tinsmiths and the traditional New Mexican santeros working in the last half of the nineteenth century might also help to identify early tinsmiths. New Mexican tin pieces have been found at Acoma, Isleta, Jemez, San Ildefonso, Santa Clara, and Taos pueblos. Additional study of the acceptance and use of tin devotional objects by the Pueblo Indians could reveal information about the Indians' relationships with their Spanish neighbors and with the Catholic church.

We are disappointed that so few pieces can be associated with individual craftsmen and hope that additional research on the specific workshops will begin to fill in these names and reveal additional workshops. For example, determining early twentieth-century working dates for the Isleta Tinsmith offers a chance to learn the identity of this craftsman. His children or grandchildren may still be living and eventually could be located if they are still in the state.

Dating New Mexican tinwork is still problematic. Accurate measurements of the thickness of the tin coatings and determination of the type of processing of the iron and steel sheets might help identify very early tin pieces. Further study of the European steel engravings and lithographs would make the dating of New Mexican tinwork more accurate.

We realize that this initial study of tinwork is not definitive. In the same manner that fifty years of ongoing research has unearthed and altered the names and identities of New Mexican santeros, the names of workshops in our study will likely change as more individual craftspeople are identified. Tinsmiths eventually may be connected to specific works as a result of the location of pieces that remain in the possession of their descendants. Some of the workshops represented by more than one style of work may be split into two or more groups and be renamed. We welcome any further scholarship that would expand and further clarify the history of New Mexican tinwork.

▼▼▼▼▼▼▼▼▼▼▼▼
NOTES

INTRODUCTION

1. Dorothy S. Kendall, *Gentilz: Artist of the Old Southwest*, pp. 59, 90–91.

2. See Sarah Nestor, *The Native Market*, for a discussion of these artists' involvement in Spanish and Indian crafts, and Charles C. Eldredge, Julie Schimmel, and William H. Truettner, *Art in New Mexico*, pp. 101–16, for a similar discussion concerning architecture and crafts.

3. It is interesting to note that E. Boyd made some of the watercolor renderings for *The Index of American Design* project.

4. The photographs are in the collections of the Photographic Archives of the Museum of New Mexico. Mr. Mathews was also an advisor on Spanish craft design to the Native Market in the late 1930s and a serious amateur photographer.

CHAPTER I

1. Eleanor B. Adams and Fray Angelico Chavez, *The Missions of New Mexico, 1776*, references throughout.

2. Hal Cannon, *Utah Folk Art*, p. 79. Quoted from "First General Epistle to the Latter-Day Saints," p. 230.

3. David J. Weber, ed., *The Extrajaneros*, pp. 20–23.

4. Lansing Bloom, ed., *Antonio Barreiro's Ojeada Sobre Nuevo Mexico*, p. 24. Barreiro, a lawyer sent to New

Mexico by the Mexican government, was asked to submit a report on the country and the population.

5. Eugene T. Wells, "The Growth of Independence, Missouri 1827–1850," *Bulletin of the Missouri Historical Society* 16, pp. 33–46.

6. John Francis McDermott, ed., *Travels in Search of the Elephant*, p. 127.

7. W.H. Emory, *Notes of a Military Reconnaissance . . .*, p. 34.

8. Ralph P. Bieber, ed., *Marching with the Army of the West, 1846–1848*, p. 163; Ibid., pp. 39–40; Waugh, p. 124.

9. *Santa Fe Republican*, January 1, 1847.

10. E. Boyd, *Popular Arts of Spanish New Mexico*, pp. 295–96.

11. One of these tinsmiths, Luis Alarid, lived next door to the priest of the Guadalupe church. It is not known whether the tinsmith's proximity indicated the use of tin for religious objects or was merely a coincidence.

12. John Baptist Salpointe, *Soldiers of the Cross*, pp. 194–95.

13. W.W.H. Davis, *El Gringo or, New Mexico and Her People*, p. 49. Davis was the U.S. Attorney for New Mexico, circa 1853–56, and editor of the *Santa Fe Weekly Gazette*, 1854–56. This is the same church described by Waugh in 1846.

14. Ibid., p. 223.

15. Cleofas M. Jaramillo, *Shadows of the Past*, p. 41.

16. Two established firms were the Spiegelberg Brothers of Santa Fe and the Huning Company of Albuquerque. Solomon Jacob Spiegelberg came to New Mexico with the first U.S. Army troops in 1846 and established his trading firm the same year. He was soon joined by two of his brothers and sometime later by the three youngest brothers. Although Solomon Jacob Spiegelberg left New Mexico in 1854, the firm of Spiegelberg Brothers was one of the leading wholesale houses of the territory throughout the nineteenth century. (Floyd S. Fierman, *Merchant-Bankers of Early Santa Fe 1844–1893*, pp. 11–20.)

Franz Huning came to New Mexico from Germany in 1849. After working for a trader in San Miguel, he moved to Albuquerque in 1852 and became a clerk. In 1857, he started a small store on the plaza in Albuquerque, where he was joined by his brother Charles. The Hunings began to drive their own goods directly from the suppliers in St. Louis rather than buying through wholesalers in Santa Fe. During the Civil War, they established the first of many branch stores at Zuni Pueblo in western New Mexico. The Huning Mercantile Company is still in operation in Los Lunas. Other European-born merchants and wholesalers of the period include the firm of Zadoc and Abraham Stabb started in 1858, the Seligman Brothers' firm established in Santa Fe in 1856, and the Ilfeld Company of Las Vegas and Santa Fe. (Franz Huning, *Trader on the Santa Fe Trail*.)

17. William Ritch, *Santa Fe: Ancient and Modern*, p. 30.

18. Jim Sagel, ed., *La Iglesia de Santa Cruz de la Cañada*, p. 21.

19. *Santa Fe Daily New Mexican*, May 24, 1889. The firm of Wagner and Hoffner advertised "both picture frames and mouldings."

20. Three Hispanic tinsmiths are listed in the census in 1900, and by 1910 two of these were listed as plumber/tinners.

21. L. Bradford Prince, *Spanish Mission Churches of New Mexico*, p. 195.

CHAPTER 2

1. Katherine M. McClinton, *The Chromolithography of Louis Prang*, p. 125.

2. *Santa Fe Daily New Mexican*, August 22, 1874.

3. United States Census, *Products of Industry Schedule*, Santa Fe County, 1880.

4. Two nineteenth-century tin pieces have been found bearing an extremely unusual stamp. The stamp, which was commercially made and is quite large—1½ inches long and ¾ inch across, is in an elongated oak-leaf form with several large lobes. The source and original purpose

of the tool are unknown, but it may have been used by a saddler. An example of a tin frame that shows this stamp is included in the Mathews photographs in the Photo Archives of the Museum of New Mexico, Neg. No. 21486.

5. Tinplate was first developed during the fourteenth century in Germany on hand-hammered iron sheets. Germany remained the principal source of tinplate until about 1750, when the British tinplate industry successfully developed a cheaper and better-quality product with the development of Pontypool, or hot-rolled iron sheets, about 1700 and the discovery of an inexpensive method of acid etching the iron sheets about 1730. British tinplate was exported as early as 1750; by 1800 Britain had become the main source of the world's tinplate, and it remained so for most of the nineteenth century.

6. W.E. Minchinton, *The British Tin Plate Industry, A History*, pp. 1–16. Tin cans were a rather recent development. Both the British in 1780 and Napoleon at the beginning of the nineteenth century offered prizes for the invention of improved ways to preserve foods. The French prize was won by a Frenchman, Appert, in 1809; his published methods called for glass containers. The first patent for a tin container was issued in England in 1812 and in the United States in 1825 (ibid., pp. 254–56). Early English canning experiments in the 1820s were done primarily for military and for Arctic expeditions. Tin cans finally replaced glass containers in America in the late 1830s, and by 1840 canned foods, including oysters and fish, were available for civilian consumption. Through the 1840s and 1850s, however, tinned foods were not widely accepted or trusted by the public. The American Civil War, along with Gail Borden's condensed milk, were responsible for the widespread acceptance of canned foods and the expansion of the industry in the first half of the 1860s (ibid., p. 258). The American meat-packing industry had its start in Chicago during this period.

7. Diana S. Waite, *Nineteenth Century Tin Roofing and Its Use at Hyde Hall*, p. 14.

8. Ibid., p. 22.

9. The traders, St. Vrain and Bent, advertised lead castings "suitable for the army or family use." *Santa Fe Republican*, October 23, 1847.

10. E. Boyd, *Popular Arts of Spanish New Mexico*, p. 196, and *Antiques* 44 (August 1943): 58.

11. A number of old tin frames have been damaged by the extra weight of double-thick plate-glass mirrors which have been added to make the frames more useful to collectors.

12. *Santa Fe Republican*, October 1, 1847. B.F. Coons advertised "paints and dyestuffs," and in June 1848, "paints and oils."

CHAPTER 3

1. E. Boyd, *Popular Arts of Spanish New Mexico*, p. 315.

2. Ibid., p. 321.

3. Carmen Espinosa, *New Mexico Tin Craft*, foreword, p. 1.

4. United States Patents 1863,001; 1864-003, microfilm copies. G.W. Prince patented an "improvement in the manufacture of tin cans" on June 28, 1864. The patent was for methods of constructing a square oil can that would withstand rough handling because of strengthened solder joints. Herman Miller's patent of 1863 was also concerned with methods of joining the top and bottom of the can in order to make it easier to ship. Such improvements, brought on by the needs of the military during the Civil War, were helpful in the rough journey along the Santa Fe Trail.

5. *Santa Fe Daily Democrat*, January 31, 1882.

6. G.K. Renner, "The Kansas City Meat Packing Industry Before 1900," *Missouri Historical Review* 56 (October 1961): 18–29.

7. Hyla M. Clark, *The Tin Can Book*, pp. 27–28.

8. Denver City Directories, 1885–1910.

9. *The Santa Fe Daily New Mexican*, October 12, 1889–March 12, 1891.

10. Yvonne Lange, "Lithography, An Agent of Technological Change in Religious Folk Art: A Thesis," *Western Folklore* 33: 52.

11. Beatrice Farwell, *French Popular Lithographic Imagery 1815–1870*, Vol. I.

12. Catherine Rosenbaum-Dondaine, *L'image de Piété en France—1814–1914*, pp. 13–14.

13. Currier and Ives, *A Catalogue Raisonné*, p. xvi.

14. Harry T. Peters, *American on Stone, The Other Printmakers to the American People*, p. 259. Currier and Ives prints can be dated from 1838 to 1872 at 152 Nassau Street; 1874 to 1877 at 123 Nassau Street; and 1877 to 1894 at 115 Nassau Street.

15. Ibid., p. 390 and p. 201.

16. *Santa Fe Daily New Mexican*, September 21, 1872.

CHAPTER 4

1. See Bibliography for Waugh, Gibson, Edwards, and others.

2. John Galvin, ed., *The Original Travel Diary of Lieutenant J.W. Abert*, p. 39.

3. Eleanor B. Adams and Fray Angelico Chavez, *The Missions of New Mexico*, 1976, references throughout.

4. Katherine S. Howe and David Warren, *The Gothic Revival Style in America 1830–1870*, details throughout.

5. José Raphael Aragón, working dates circa 1820–62. See William Wroth, Christian Images in Hispanic New Mexico, pp. 129–31.

6. In Mexico these chandeliers are referred to as *coronas*.

7. Lansing B. Bloom, "Bourke on the Southwest," *New Mexico Historical Review* 11 (3): 250.

8. W.W.H. Davis, *El Gringo or, New Mexico and Her People*, p. 49.

9. Marc Simmons and Frank Turley, *Southwestern Colonial Ironwork*, pp. 88–89.

10. Taylor Museum Accession Record.

11. Personal communication, Will Wroth to Maurice Dixon, March 29, 1987.

12. *Index of American Design Worksheets*, NM ME-57. The torches were owned in 1937 by George Travis, an antique dealer in Taos.

13. A similar candle mold has been attributed to Christopher Riding, an English tinsmith working in St. George, Utah, about 1870. See Hal Cannon, *Utah Folk Art*, p. 84.

14. Personal communication, Lonn Taylor to Lane Coulter, March 5, 1987.

CHAPTER 5

1. Manuscript Census Returns, The United States Census, on file in the New Mexico State Records Center and Archives, Santa Fe, New Mexico. See Appendix.

2. Cleofas M. Jaramillo, *Shadows of the Past*, p. 41. For a further discussion of José María Baca, see the section on the Rio Arriba Workshop in this chapter.

3. E. Boyd, *Popular Arts of Spanish New Mexico*, Figure 176, p. 298.

4. Museum of International Folk Art Accession Records.

5. See Figure 1.1 for another example of a datable piece from this workshop.

6. Jim Sagel, ed., *La Iglesia de Santa Cruz de las Cañada*, 1983.

7. This piece was formerly in the collection of the late Randall Davey, an early Santa Fe painter.

8. Opal, Harber, *Photographers and the Colorado Scene 1855–1900*, n.p.

9. According to her biographer and the current owner of the home, Miss Hollenback kept extensive records of her acquisitions. These sconces had old tags attached to the back at the time they were acquired.

10. Jerry L. Williams and Paul E. McAllister, eds., *New Mexico in Maps*, pp. 41 and 53.

11. Cleofas M. Jaramillo, *Shadows of the Past*, p. 41. Note: Mrs. Jaramillo was born in the 1870s, and her recollections must surely date from as early as 1885.

12. Personal Communication, Fr. Jerome Martínez y Alire to Lane Coulter, April 28, 1987.

13. Marianne Stoller, *A Study of Nineteenth Century Hispanic Arts and Crafts in the American Southwest: Appearances and Processes*, pp. 519–20.

14. Richard E. Ahlborn, *The Penitente Moradas of Abiquiu*, p. 162.

15. Museum of International Folk Art Accession Records: Nicho with lunette containing the fourteen apparitions of St. James Major in the form of miniature bultos. The large nicho is 31 inches tall by 14 inches wide by 15 inches deep. See E. Boyd, 1959, p. 49, for an illustration.

16. Two almost identical nichos are illustrated in Christine Mather, ed., *Colonial Frontiers*, p. 5.

17. David Bell, *Albuquerque Journal North*, July 29, 1986, p. 4.

18. Manuscript Census Returns, Mora County, 1880.

19. Jean Stern, ed., *The Cross and the Sword*, p. 103, and Norman Neuerburg, *Saints for the People*, p. 63.

20. Southwest Museum Accession Records, and *Index of American Design Work Sheets*, Southern California project.

21. Interview with Ben Apodaca Martínez, grandson of José María Apodaca, August 6, 1988.

22. Ibid.

23. Manuscript Census Returns, Santa Fe County, 1910.

24. This particular type of frame was a favorite of early collectors, who replaced the religious images with mirrors. Sometimes they commissioned Revival tinsmiths to convert the frames to electrified sconces.

25. A line drawing of a similar frame (photographed in the 1930s by Truman Mathews) is included in the 1937 vocational workbook (Carmen Espinosa, *New Mexico Tin Craft*, p. 14). At that time the frame was in the collection of Mary Wheelwright of Alcalde, New Mexico, although its present location is unknown. This same frame also served as a model for two light fixtures made during the Revival period for the McKinley County Courthouse in Gallup, New Mexico.

26. See Photo Archives, Museum of New Mexico, Neg. No. 21408, for Mathews photograph.

27. Colorado Historical Society #H.6000.137.

28. Several distinctly architectural frames are included in the Mathews photographs. See Photo Archives, Museum of New Mexico.

29. Espinosa, *New Mexico Tin Craft*, p. 8.

30. John Adair, *The Navajo and Pueblo Silversmiths*, p. 184.

31. Patrick T. and Betsy E. Houlihan, *Lummis in the Pueblos*, pp. 31–34, 49.

32. *New Mexico State Directories*, 1906, 1912–16. Mr. August Seis assumed management of a general store in Isleta from his father, George H. Seis, about 1912.

33. Personal interview, Tony Garcia with Lane Coulter, October 21, 1986, and Dr. Ted Jojola with Lane Coulter, April 7, 1987.

34. Adair, *The Navajo and Pueblo Silversmiths*, pp. 180–83.

35. Personal communication, Tony Garcia to Lane Coulter, November 2, 1988.

36. Photo Archives, Museum of New Mexico, Neg. No. 21489.

37. E. Boyd, *Popular Arts of Spanish New Mexico*, p. 432.

38. This company still produces terneplate in West Virginia.

39. For a complete discussion of this work, see James S. Griffith, "The Magdalena Holy Pictures: Religious Folk Art in Two Cultures," *New York Folklore* 8, nos. 3 and 4 (Winter 1982): 71–82.

40. Ibid., p. 77.

41. Letter, James S. Griffith to Lane Coulter, March 14, 1987.

42. Patrick T. and Betsy E. Houlihan, *Lummis in the Pueblos*, p. 27.

43. Faustin Vigil was born in 1834 and is listed in the Mexican Census of 1845 for the Province of New Mexico, the Santa Clara Pueblo area (State Records Center and Archives, Mexican Archives).

44. The nicho in Figure 5.104 was collected before 1931 by Santa Fe painter and collector Frank Applegate.

45. This nicho was illustrated in the magazine *Antiques* (August 1943): 61. In the article the nicho is dated circa 1830–40, although it was probably made about 1860. The drawing and the use of glass panels would suggest that this date is too early.

46. For an additional example of these watercolors, see Figure 2.3.

47. Personal interview, J. Paul Taylor with Lane Coulter, October 16, 1986.

CHAPTER 6

1. Charles C. Eldredge, Julie Schimmel, and William H. Truettner, *Art in New Mexico, 1900–1945*, p. 104.

2. David Gebhard, "Architecture and the Fred Harvey Houses," *New Mexico Architect* 4 (7 and 8): 11–17, and 6 (1 and 2): 18–25.

3. Virginia L. Grattan, *Mary Colter: Builder Upon the Red Earth*, p. 125.

4. John L. Kessell, *The Missions of New Mexico Since 1776*, pp. 24–26.

5. Ibid., p. 33.

6. Rosemary Nussbaum, *Tierra Dulce, Reminiscences from the Jesse Nussbaum Papers*, pp. 61–64.

7. Eldredge, Schimmel, and Truettner, *Art in New Mexico, 1900–1945*, p. 112.

8. *El Palacio* 18, no. 7 (April 1925): 1.

9. Ibid., and New Mexico State Business Directory, 1925.

10. Personal interview, Pedro Quintana with Lane Coulter, June 6, 1987. For a more complete description of Mr. Quintana's background see Sarah Nestor, *The Native Market of the Spanish New Mexican Craftsmen*, p. 25.

11. Ibid., p. 52.

12. Ibid., p. 6.

13. William Wroth, "New Hope in Hard Times," *El Palacio* 89 (2): 27.

14. William Wroth, *Hispanic Crafts of the Southwest*, p. 105, and U.S. Census Returns, Taos County, 1860–1910.

15. Personal interview, Virginia Hunter Ewing with Lane Coulter, July 8, 1987.

16. U.S. Census Returns, Santa Fe County, 1910; Santa Fe City Directories, 1925–45; Carmen Espinosa, *New Mexico Tin Craft*, foreword.

17. Francisco Sandoval's father, José Sandoval, was a tinsmith in the mid-nineteenth century, according to family tradition. Bonafacio Sandoval to Lane Coulter, July 15, 1987.

18. Personal interview, Pedro Quintana with Lane Coulter, June 5, 1987.

19. Santa Fe City Directories, 1925–36, and "Eddie Delgado—Creative Artisan," *Santa Fe Scene* 1, no. 8 (March 8, 1958): 5.

20. *Santa Fe City Directories*, 1925–45.

21. *Art and Archaeology* 18, nos. 5 and 6 (December 1924): 240.

22. *School Arts* 33, no. 9 (May 1934): 529.

23. Nestor, *The Native Market*, p. 29.

24. Personal interview, Pedro Quintana with Lane Coulter, June 5, 1987.

25. This design is a common representation of the Trinity. Personal communication, Gloria Giffords to Lane Coulter, December 1987.

26. Personal interview, Quintana with Coulter, June 5, 1987.

27. *El Palacio* 23 (12): 337.

CHAPTER 7

1. Interviews with Jack Caldarella, El Paso, Texas, July 23, 1986 and December 28, 1987; Robin Cleaver, Santa Fe, New Mexico, October 12, 1987.

2. W.E. Minchinton, *The British Tinplate Industry, A History*, p. 15.

3. Eleanor B. Adams and Fray Angelico Chavez, *The Missions of New Mexico, 1776*, references throughout.

4. Ibid., p. 65. Father Tagle was in residence at San Ildefonso by 1715.

5. Gloria K. Giffords, *Mexican Folk Retablos*, p. 22.

6. Ibid., p. 123.

7. Personal communication, Gloria Giffords to Lane Coulter, June 8, 1987.

8. Richard E. Ahlborn, *The Penitente Moradas of Abiquiu*, p. 154.

GLOSSARY

BOBECHE   Shallow dish portion of a candle holder used to catch candle-wax drippings.

BOSS   Raised design detail, generally round, often terminating a linear element.

BULTO   Three-dimensional religious statue, usually of carved wood or cast plaster.

CHROMOLITHOGRAPH   Multicolored print produced by the lithographic process.

DIAPER   Two-dimensional design pattern with the image of a grid of diamond shapes.

DIE   Punching tool used for striking an image.

EMBOSSING   Process of striking an image with a die from the reverse (see STAMPING).

ENGRAVING   Printmaking technique in which a design is cut into a metal sheet, and a print made from the inked metal plate. Copper-plate and steel-plate engravings were the most common.

FINIAL   Ornamental terminal at the top of a form.

FLUTED   Decorated with a detail consisting of parallel grooves.

FLUX   Material used in the soldering process to clean the metal and to prevent oxidation during heating.

LITHOGRAPH   Print produced by process of lithography in which the image is drawn on a stone and treated so that it retains oil-based inks and the nonimage areas repel ink.

LUNETTE   Semicircular or crescent-shaped design element attached to the top of a form.

MORADA   Chapel and chapter house of the Penitente brotherhood.

NICHO   Niche; religious structure made to contain a two- or three-dimensional holy image.

OLEOGRAPH   Chromolithograph made in imitation of an oil painting. The surface is usually varnished.

PEDIMENT   Design element consisting of a wide triangular shape mounted atop a square or rectangle.

PENITENTE   Lay religious society of the Catholic Church in the American Southwest based on the Passion of Jesus and the spirit of penance. Also known as *Los Hermanos Penitentes* (the Penitente brotherhood).

PUNCHING   Process of striking an image into a sheet material with a die. Can be either stamped or embossed.

QUATREFOIL   Four-lobed decorative design element.

RELIQUARY   Religious shrine for displaying sacred relics.

REPOUSSÉ   Metalworking process in which design areas are raised from the reverse of a metal sheet with repeated hammering.

RETABLO   Two-dimensional painting of a religious image.

ROSIN   Material derived from pine-tree sap used in soldering flux.

SOLOMONIC or SALOMONICA   Spiral-twisted column used as a design element.

SANTERO   Maker of saints and religious images.

SANTO   Any religious image, including a bulto, retablo, or print.

SCONCE   Wall-mounted candle holder.

SCORING   Metalworking process used to create a continuous line.

SOLDER   Metallic material used to join sheets of metal. In tinsmithing a lead-tin alloy is used.

STAMPING   Process of striking an image from the front of a metal sheet with a die.

TERNE, TERNEPLATE   Iron or steel sheet coated with a tin-lead alloy.

TINPLATE   Iron or steel sheet coated with tin. Products made from tinplate.

TREFOIL   Three-lobed decorative design element.

WOODCUT   Print made by cutting an image in reverse into a wooden block, then printing it on paper.

▼▼▼▼▼▼▼▼▼▼▼▼▼▼▼▼▼▼▼

APPENDIX

Hispanic Tinsmiths in the
United States Census, 1850–1910,
and Other Hispanic Tinsmiths

1850

Santa Fe County (*City of Santa Fe*)
Luis Alarid, age 22, born NM, tinner
Miguel Maes, age 35, b. NM, tinner
Juan Maes, age 16, b. NM, tinner
Jesús María Martín(ez), age 22, b. NM, tinner
José María Martín(ez), age 62, b. NM, tinner
Ygnacio Valdes, age 27, b. NM, listed as a hatter

1860

Santa Fe County
(*City of Santa Fe*)
Miguel Maes, age 43, b. NM, tinner
(illegible), age 28, b. Mexico, tinsmith
Luis __i _a(?), age 37, b. NM, tinner
(*Cuyamungue*)
Ygnacio Valdes, age 38, b. NM, tinner

1870

Mora County
Juan Espinosa, age 25, b. NM, tinsmith
Santa Fe County
(*City of Santa Fe*)
Alofia Avila, age 36, b. NM, tinworker
(*San Ildefonso*)
Ejinia [sic] (Higinio) Gonzales, age 28, b. NM, tinner

1880

Bernalillo County (*Albuquerque*)
Horacio Quiñones, age 20, b. Chihuahua, Mexico, tinsmith
Francisco Ruiz, age 58, b. Chihuahua, Mexico, tinsmith
Mora County (*La Cueva*)
Juan Armijo, age 56, b. NM, tinsmith
San Miguel County (*La Cuesta*)
Ricardo Gallegos, age 37, b. NM, tinner
Santa Fe County
(*City of Santa Fe*)
Gavino Borrego, age 50, b. NM, tinner
Juan J. Coca, age 64, b. NM, tinner
Francisco Ortíz, age 16, b. NM, tinner
Luis Ortíz, age 22, b. NM, tinsmith
Felix Quintana, age 15, b. NM, tinner
Antonio Velocia, age 46, b. NM, tinner
(*Cuyamungue*)
Ygnacio Valdez, age 58, b. NM, listed as a laborer

1890 (*census returns destroyed*)

1900

Santa Fe County
(*City of Santa Fe*)
Francisco Ortíz y Baca, age 38, b. NM, listed as jeweler
Luis Ortíz, age 51?, b. NM, tinsmith
Felix Quintana, age 34, b. NM, tinsmith
(*Ojo de la Vaca*)
José María Apodaca, age 55, b. NM?, no occupation listed
Valencia County (*Belen*)
Francisco Quiroz, age 42, b. Old Mexico (moved to NM in 1885), tinner

1910

Santa Fe County
(*City of Santa Fe*)
Louis Ortíz y Baca, age 52, b. NM, plumber/tinner
Felix Quintana, age 46, b. NM, plumber in tin shop
Francisco Sandoval, age 23?, b. NM, owns tin shop
(*Ojo de la Vaca*)
José María Apodaca, age 64, b. NM?, farmer

OTHER HISPANIC TINSMITHS

José María Baca, Abiquiú, c. 1880 (from Jaramillo, *Shadows of the Past*)
Felipe Garcia, Ojo de la Vaca and Santa Fe, c. 1890–1900 (family tradition)
Fred Quintano, Santa Fe (from *New Mexico Pictorial State Gazetteer*), 1912–13
Teodocio Ortíz, Nambé (from *New Mexico State Directory 1915*)

BIBLIOGRAPHY

UNPUBLISHED MATERIALS

Batchen, Lou Sage. "Life in the Old Houses: Los Pedlers." Typed manuscript, 1937. WPA Files, History Library, Museum of New Mexico.

City of Santa Fe Retail Merchant License Receipts 1862–92. On file in the New Mexico State Records Center and Archives, Santa Fe.

Heath, Jim F. *A Study of the Influence of the Atchison, Topeka, and Santa Fe Railroad on the Economy of New Mexico, 1878 to 1900.* Master's thesis, University of New Mexico, 1955.

Hunter, R. Vernon. "Spanish Colonial Arts." Typed manuscript, c. 1937. WPA Files, History Library, Museum of New Mexico.

Ilfeld Family Papers, 1879–1930. New Mexico State Records Center and Archives, Santa Fe.

Index of American Design Work Sheets. On file, National Gallery of Art, Washington, D.C.

Mexican Census of 1845 for the Province of New Mexico. On file, Mexican Archives, New Mexico State Records Center and Archives, Santa Fe.

Spiegelberg, Flora. Letters to Hester Jones and Paul A.F. Walter, October 17, 1933 and January 13, 1935. New Mexico State Records Center and Archives, Santa Fe.

United States Manuscript Census Returns, 1850–1910. Microfilm in New Mexico State Records Center and Archives, Santa Fe.

NEWSPAPERS AND DIRECTORIES

Colorado, New Mexico, Nevada, Wyoming, and Arizona Gazeteer and Business Directory—1884–85. Chicago, Detroit, and St. Louis: R.L. Polk and Co., 1884.

Denver City Directories, 1885–1910.

McKenny's Business Directory, 1882–83 and 1888–89. Oakland, California: Pacific Press, 1882.

New Mexico and Arizona Pictorial Gazeteer and Business Directory. St. Paul: R.L. Polk and Co., 1912–13.

New Mexico State Business Directory. Denver: Gazeteer Publishing Co., 1905–6, 1913–32, 1936, 1938, 1943.

Santa Fe City Directory. El Paso: Hudspeth Directory Co., 1924–44.

Various newspapers, New Mexico, 1847–1900. Microfilm on file, New Mexico State Records Center and Archives, Museum of New Mexico History Library, Santa Fe.

INTERVIEWS

Caldarella, Jack. El Paso, Texas, July 23, 1986, and December 28, 1987.

Cash, Marie Romero. Santa Fe, New Mexico, December 9, 1988.

Cleaver, Robin. Santa Fe, New Mexico, October 12, 1987.

Cook, Mary Jean. Santa Fe, New Mexico, October 8, 1987.

Espinosa, Carmen. Albuquerque, New Mexico, October 8, 1986.

Ewing, Virginia Hunter. Santa Fe, New Mexico, July 8, 1987.

Garcia, Tony. Alameda, New Mexico, October 21, 1986.

Giffords, Gloria K. Tucson, Arizona, June 8, 1987.

Herrera, Raymond M. Santa Fe, New Mexico, August 4, 1988.

Jojola, Ted. Albuquerque, New Mexico, April 7, 1987.

Lange, Yvonne. Santa Fe, New Mexico, April 14, 1987.

Martínez, Ben Apodaca. Santa Fe, New Mexico, August 6, 1988.

Martínez y Alire, Fr. Jerome. Albuquerque, New Mexico, April 28, 1987.

Quintana, Pedro. Santa Fe, New Mexico, June 5, 1987.

Romero, Emilio. Santa Fe, New Mexico, July 15, 1987.

Sandoval, Bonafacio. Albuquerque, New Mexico, January 6, 1986, and July 15, 1987.

Taylor, J. Paul. Mesilla, New Mexico, October 16, 1986.

Taylor, Lonn. Washington, D.C., March 5, 1987.

BOOKS AND ARTICLES

Abert, J.W. *Report on His Examination of New Mexico in the Years 1846–1847*. Washington, D.C.: Senate Executive Document No. 23, Thirtieth Congress, 1st Session, Vol. IV, 1848.

Adair, John. *The Navajo and Pueblo Silversmiths*. Norman: University of Oklahoma Press, 1946.

Adams, Eleanor B., and Fray Angelico Chavez. *The Missions of New Mexico, 1776*. Albuquerque: University of New Mexico Press, 1956.

Adams, Robert H. *The Architecture and Art of Early Hispanic Colorado*. Boulder: Colorado Associated University Press with the State Historical Society of Colorado, 1974.

Ahlborn, Richard E. *The Penitente Moradas of Abiquiú*. Washington, D.C., and London: Smithsonian Institution Press, 1986.

Archdiocese of Santa Fe. *The Lord and New Mexico*. Santa Fe, 1976.

The Art of the Spanish Southwest. The United States–Mexico Commission for Border Development and Friendship, n.d. (Renderings from the *Index of American Design*, National Gallery of Art. Washington, D.C.)

Beachum, Larry M. *William Becknell, Father of the Santa Fe Trade*. El Paso: Texas Western Press, 1982.

Beck, Warren A., and Ynez D. Hasse. *Historical Atlas of New Mexico*. Norman: University of Oklahoma Press, 1969.

Bell, David. "Conversations Add Extra Pleasure to Spanish Market." *Albuquerque Journal North*, July 29, 1986.

Bieber, Ralph P., ed. *Marching with the Army of the West, 1846–1848*. Glendale: Arthur H. Clark Co., 1936.

Bloom, Lansing B. *Antonio Barreiro's Ojeada Sobre Nuevo Mexico*. Historical Society of New Mexico, Publications in History 5. Santa Fe: El Palacio Press, 1928.

———. "Bourke on the Southwest." *New Mexico Historical Review* 8–13.

Boston Museum of Fine Arts. *Frontier America: The Far West*. Boston: Museum of Fine Arts, 1974.

Boyd, E. "Antiques in New Mexico." *Antiques* 44 (August 1943): 58–62.

———. "Decorated Tinware East and West in New Mexico." *Antiques* 66 (September 1954).

——— (unsigned). *Hand List of the Spanish Colonial Arts Society, Inc.*, n.d. (c. 1954).

———. *The New Mexican Santero*. Santa Fe: Museum of New Mexico Press, 1969.

———. "New Mexican Tin Work." *El Palacio* 60, no. 2 (February 1953): 61–67.

———. *New Mexico Tinwork*. Leaflet No. 2. Santa Fe: School of American Research, 1953.

———. *Popular Arts of Colonial New Mexico*. Santa Fe: Museum of International Folk Art, 1959.

———. *Popular Arts of Spanish New Mexico*. Santa Fe: Museum of New Mexico Press, 1974.

Brandi, Manuel Quesada, ed. *La Litografía en Mexico en el Siglo XIX*. Reprint of the 1934 edition. Mexico City: Editiones Facsimilares de las Biblioteca National de Mexico, 1965.

Bullock, Alice. *Mountain Villages*. Santa Fe: Sunstone Press, 1973.

Bunting, Bainbridge. *Early Architecture in New Mexico*. Albuquerque: University of New Mexico Press, 1976.

———. *John Gaw Meem: Southwestern Architect*. Albuquerque: University of New Mexico Press, 1983.

———. *Taos Adobes*. Santa Fe: Fort Burgwin Research Center and Museum of New Mexico Press, 1964.

Calvin, Ross. *Lt. Emory Reports*. Albuquerque: University of New Mexico Press, 1957.

Cannon, Hal. *Utah Folk Art*. Provo, Utah: Brigham Young University Press, 1980.

Cassidy, Ina Sizer. "Art and Artists in New Mexico: Home Arts and Crafts." *New Mexico* 17, no. 5 (May 1939): 26–27, 46–47.

———. "Art and Artists of New Mexico: Adventures in Tin." *New Mexico* 15, no. 8 (August 1937): 28, 56.

Charlot, Jean. *Mexican Art and the Academy of San Carlos, 1785–1915*. Austin: University of Texas Press, 1962.

Chavez, Fray Angelico. *Our Lady of the Conquest*. Santa Fe: Historical Society of New Mexico, 1948.

Clark, Hyla M. *The Tin Can Book*. New York, London, Scarborough: New American Library, 1977.

Cobos, Ruben. *A Dictionary of New Mexico and Southern Colorado Spanish*. Santa Fe: Museum of New Mexico Press, 1983.

Coffin, Margaret. *The History and Folklore of American Country Tinware 1700–1900*. New York: Galahad Books, 1968.

Comfort, Charles, ed. "Eddie Delgado—Creative Artisan." *Santa Fe Scene* 1, no. 8 (March 8, 1958): 4–7.

Comstock, Helen. *The Looking Glass in America, 1700–1825*. New York: Viking Press, 1968.

Conningham, Frederick A. *Currier and Ives Prints, An Illustrated Check List*. New York: Crown Publishers, 1970.

Currier & Ives, A Catalogue Raisonné, Vol. 1. Detroit: Gale Publishing Company, 1984.

Curtin, Leonora F. "Back to Tin—An Ancient Craft." *Touring Topics* 24, no. 9 (September 1932): 22–23.

Davis, W.W.H. *El Gringo or, New Mexico and Her People*. Santa Fe: Rydal Press, 1938.

deBorhegi, Stephen F. "The Miraculous Shrines of Our Lord of Esquípulas in Guatemala and Chimayó, New Mexico." *El Santuario de Chimayo*. Santa Fe: Spanish Colonial Art Society, 1956.

Denver Art Museum. *Santos of the Southwest*. Denver, n.d. (1970).

Dickey, Roland F. *New Mexico Village Arts*. Albuquerque: University of New Mexico Press, 1949.

Die Lithographie in Der Schweiz und die Vervandten

Techniken. Bern: Verein Schweizerischer Lithographiebesitzer, 1944.

Drumm, Stella M., ed. *Down the Santa Fe Trail and into Mexico: The Diary of Susan Shelby Magoffin, 1846–1847*. New Haven: Yale University Press, 1962.

Eldredge, Charles C., Julie Schimmel, and William H. Truettner. *Art in New Mexico, 1900–1945*. Washington, D.C., and New York: National Museum of American Art and Abbeville Press, 1986.

Ellis, Bruce T. *Bishop Lamy's Santa Fe Cathedral*. Albuquerque: University of New Mexico Press, 1985.

Ellis, Florence Hawley. "Tomé and Father J.B.R." *New Mexico Historical Review* 30, no. 2: 89–114, and no. 3: 195–220.

Emory, W.H. *Notes of a Military Reconnoissance from Fort Leavenworth, in Missouri to San Diego, in California*. Washington: Wendell and Van Benthuysen, 1848.

Espinosa, Carmen. *New Mexico Tin Craft*. Santa Fe: New Mexico State Department of Vocational Education, 1937.

Espinosa, José E. *Saints in the Valleys*. Albuquerque: University of New Mexico Press, 1967.

Farwell, Beatrice. *French Popular Lithographic Imagery, 1815–1870*. 12 vols. Chicago and London: University of Chicago Press, Vol. 1, 1981, Vols. 4–5, 1986.

Federal Art Project. *Portfolio of Spanish Colonial Design in New Mexico*. Santa Fe: Works Progress Administration, 1938.

Fierman, Floyd S. *Merchant-Bankers of Early Santa Fe*. El Paso, Texas: Western College, 1964.

"The Fiesta Art Exhibition." *El Palacio* 17, nos. 6–7 (September 1924): 169–74.

"Fiesta Park Plans." *El Palacio* 18, no. 7 (April 1925): 143–48.

Galvin, John, ed. *The Original Travel Diary of Lieutenant J.W. Abert Who Mapped New Mexico for the U.S. Army*. San Francisco: John S. Howell, 1966.

Gebhard, David. "Architecture and the Fred Harvey Houses." *New Mexico Architect* 4, nos. 7–8 (July–August 1962): 11–17, and 6, nos. 1–2 (January–February 1964): 18–25.

Gibson, George Rutledge. *Journal of a Soldier Under Kearney and Doniphan, 1846–1847*. Glendale: Arthur H. Clark Co., 1935.

Giffords, Gloria K. *Mexican Folk Retablos: Masterpieces on Tin*. Tucson: University of Arizona Press, 1974.

Grattan, Virginia L. *Mary Colter: Builder upon the Red Earth*. Flagstaff: Northland Press, 1980.

Griffith, James S. "The Magdalena Holy Pictures: Religious Folk Art in Two Cultures." *New York Folklore* 8, nos. 3–4 (Winter 1982): 71–82.

Griffiths, Therese. "Angie Delgado Martínez—Artist at Work." *New Mexico* 56, no. 1 (January 1978): 42–43.

"A group of tin lighting fixtures similar to Spanish Colonial types designed and executed by Mr. Sweringer of Santa Fe." *School Arts* 33, no. 9 (May 1934): 529.

Harber, Opal. *Photographers and the Colorado Scene, 1853–1900*. Denver: Western History Department, Denver Public Library, 1961.

Harnung, Clarence P. *Treasury of American Design*. New York: Harry N. Abrams, 1976.

Hazen-Hammond, Susan. "Art From Poor Man's Silver." *New Mexico Magazine* 62, no. 12 (December 1984): 56–61.

Horgan, Paul. *Three Centuries of Santa Fe*. New York: E.P. Dutton and Co., 1956.

Hougland, Willard. *Santos: A Primitive American Art*. New York, 1946.

Houlihan, Patrick T., and Betsy E. Houlihan. *Lummis in the Pueblos*. Flagstaff: Northland Press, 1986.

Howe, Katherine S., and David Warren. *The Gothic Revival Style in America 1830–1870*. Houston: Museum of Fine Arts, 1976.

Huning, Franz. *Trader on the Santa Fe Trail, Memoirs of Franz Huning*. Albuquerque: University of New Mexico Press in collaboration with Calvin Horn, Publisher, 1973.

Jaramillo, Cleofas M. *Shadows of the Past*. Santa Fe: Seton Village Press, 1941.

Kelly, Daniel T. *The Buffalo Head*. Santa Fe: Vergara Publishing Co., 1972.

Kendall, Dorothy S. *Gentilz: Artist of the Old Southwest*. Austin: University of Texas Press, 1974.

Kessell, John L. *The Missions of New Mexico Since 1776*. Albuquerque: University of New Mexico Press for the Cultural Properties Review Committee, 1980.

Knee, Ernest. *Santa Fe*. New York: Hastings House, 1942.

Lange, Yvonne. "Lithography, An Agent of Technological Change in Religious Folk Art: A Thesis." *Western Folklore* 33, no. 1 (January 1974): 51–64.

Lasansky, Jeanette. *To Cut, Piece, and Solder, The Work of the Rural Pennsylvania Tinsmith*. University Park: Pennsylvania State University Press—Keystone Books, 1982.

Lynn, Catherine. *Wallpaper in America*. New York: W.W. Norton, 1980.

Mather, Christine. "The Arts of the Spanish in New Mexico." *Antiques* 113 (February 1978): 422–29.

———. *Baroque to Folk*. Santa Fe: Museum of New Mexico Press for the Museum of International Folk Art, 1980.

———, ed. *Colonial Frontiers*. Santa Fe: Ancient City Press for the Museum of International Folk Art, 1983.

Mauzy, Wayne. "Santa Fe's Native Market." *El Palacio* 40, nos. 13–15 (March–April 1936): 65–72.

McClinton, Katherine M. *The Chromolithography of Louis Prang*. New York: Clarkson N. Potter, 1973.

McDermott, John Francis, ed. *Travels in Search of the Elephant: The Wanderings of Alfred S. Waugh, Artist, in Louisiana, Missouri, and Santa Fe, in 1845–46*. St. Louis: Missouri Historical Society, 1951.

McKie, James W. *Tin Cans and Tin Plate*. Cambridge, Mass.: Harvard University Press, 1959.

Mills, George. *The People of the Saints*. Colorado Springs: Taylor Museum of the Colorado Springs Fine Arts Center, 1967.

Minchinton, W.E. *The British Tinplate Industry, A History*. Oxford: Clarendon Press, 1957.

Moorhead, Max. *New Mexico's Royal Road: Trade and Travel on the Chihuahua Trail*. Norman: University of Oklahoma Press, 1958.

"Museum Events: Spanish Colonial Arts." *El Palacio* 23, no. 12 (September 1927): 337–39.

National Endowment for the Arts. "The National Heritage Fellowships 1987." Washington, D.C.: National Endowment for the Arts Folk Arts Program, 1987.

Nestor, Sarah. *The Native Market of the Spanish New Mexican Craftsmen, Santa Fe, 1933–1940*. Santa Fe: Colonial New Mexico Historical Foundation, 1978.

Neuerburg, Norman. *Saints for the People*. Los Angeles: Loyola Marymount University, 1982.

Nussbaum, Rosemary. *Tierra Dulce, Reminiscences from the Jesse Nussbaum Papers*. Santa Fe: Sunstone Press, 1980.

Nylander, Richard C. *Wallpapers for Historic Buildings*. Washington, D.C.: Preservation Press, 1983.

Oman, Charles C., and Jean Hamilton. *Wallpapers: An International History and Illustrated Survey from the Victoria and Albert Museum*. New York: Harry N. Abrams, 1982.

Pearce, T.M., ed. *New Mexico Place Names, A Geographical Dictionary*. Albuquerque: University of New Mexico Press, 1965.

Peters, Harry T. *America on Stone, The Other Printmakers to the American People*. Garden City and New York: Doubleday, Doran, and Co., 1931.

———. *Currier and Ives, Printmakers to the American People*. Garden City and New York: Doubleday, Doran and Co., 1942.

Prince, L. Bradford. *Spanish Mission Churches of New Mexico*. Rapid City, Iowa: Torch Press, 1915.

Renner, G.K. "The Kansas City Meat Packing Industry Before 1900." *Missouri Historical Review* 56 (October 1961): 18–29.

The Resources of New Mexico—1881. Reprint of the 1881 edition for the Territorial Fair. Santa Fe: William Gannon, 1973.

Ritch, William. *Santa Fe: Ancient and Modern*. Santa Fe: Bureau of Immigration, 1885.

Robacker, Earl F. "Decorated Tinware East and West in Pennsylvania." *Antiques* 66 (September 1954).

Robertson, Edna, and Sarah Nestor. *Artists of the Canyons and Caminos. Santa Fe, The Early Years.* Salt Lake City: Peregrine Smith, 1982.

Roche, Serge. *Mirrors.* London: Gerald Duckworth and Co., 1957.

Rosenbaum-Dondaine, Catherine. *L'image de Pieté en France—1814–1914.* Paris: Musée-Galerie de la Seita, 1984.

Ross, Patricia. "The Craft of Tinwork." *Contemporary Arts of the Southwest* I, no. 2 (January–February 1933): 11.

Rusinow, Irving. *A Camera Report on El Cerrito: A Typical Spanish-American Community in New Mexico.* Washington, D.C.: U.S. Department of Agriculture, 1942.

Sagel, Jim, ed. *La Iglesia de Santa Cruz de la Cañada.* Santa Cruz: Holy Cross Parish, 1983.

———. "Romeros Continue Artist Tradition Begun by Emilio Sr." *Albuquerque Journal North*, July 25, 1984.

Salpointe, John Baptist. *Soldiers of the Cross.* Albuquerque: Calvin Horn, Publisher, 1967.

Schiffer, Herbert F. *The Mirror Book, English, American and European.* Exton, Penn.: Schiffer Publishing, 1983.

Shalkop, Robert L. *Arroyo Hondo: The Folk Art of a New Mexican Village.* Colorado Springs: Taylor Museum of the Colorado Springs Fine Arts Center, 1967.

Sheppard, Carl D. *Creator of the Santa Fe Style.* Albuquerque: University of New Mexico Press, 1988.

Shipway, Vera Cook, and Warren Shipway. *The Mexican House: Old and New.* New York: Architectural Book Publishing Co., 1960.

Shockley, Linda. "Celebrating and Collecting Spanish Culture." *Pasatiempo* (July 25, 1986): 4.

Simmons, Marc, and Frank Turley. *Southwestern Colonial Ironwork.* Santa Fe: Museum of New Mexico Press, 1980.

Steele, Thomas J. *Santos and Saints.* Santa Fe: Ancient City Press, 1974.

Stern, Jean, ed. *The Cross and the Sword.* San Diego: Fine Arts Society of San Diego, 1976.

Stoller, Marianne L. *A Study of Nineteenth Century Hispanic Arts and Crafts in the American Southwest: Appearances and Processes.* Ph.D. dissertation, University of Pennsylvania, 1979. Ann Arbor: University Microfilms, n.d.

Sunder, John E., ed. *Matt Field on the Santa Fe Trail.* Norman: University of Oklahoma Press, 1960.

Tate, Bill. *The Penitentes of the Sangre de Cristos.* Truchas, N.M.: Tate Gallery, 1968.

Taylor, Lonn, and Dessa Bokides. *New Mexican Furniture, 1600–1940.* Santa Fe: Museum of New Mexico Press, 1987.

Toro, Alphonso. "The Art of Engraving in Mexico." *Mexican Art and Life* 5 (January 1939).

Treasury Department. *The Foreign Commerce and Navigation for the Year Ending June 30, 1893.* Washington, D.C.: Government Printing Office, 1894.

Tucson Museum of Art. *Imágenes Hispanoamericanas.* Tucson: Tucson Museum of Art, 1976.

Van Stone, Mary R. "The Fiesta Art Exhibition." *Art and Archaeology* 18, nos. 5–6 (December 1924): 225–40.

Von Wuthenau, Alexander. "The Spanish Military Chapels in Santa Fe and the Reredos of Our Lady of Light." *New Mexico Historical Review* 10 (1935): 175–94.

Waite, Diana S. *Nineteenth Century Tin Roofing and Its Use at Hyde Hall.* Albany: New York State Division for Historic Preservation, 1974.

Warner, Louis Henry. *Archbishop Lamy; An Epoch Maker.* Santa Fe: Santa Fe New Mexican Publishing Corp., 1936.

Weaver, Warren A. *Lithographs of N. Currier and Currier and Ives.* New York: Holport Publishing Co., 1925.

Webb, James Josiah. *Adventures in the Santa Fe Trade 1844–1847.* Glendale, Calif.: Arthur H. Clark Co., 1931.

Weber, David J., ed. *The Extranjeros: Selected Documents from the Mexican Side of the Santa Fe Trail, 1825–1828.* Santa Fe: Stagecoach Press, 1967.

Weigle, Marta. *Brothers of Light, Brothers of Blood*. Albuquerque: University of New Mexico Press, 1976.

———, ed. *Hispanic Arts and Ethnohistory in the Southwest*. Santa Fe: Ancient City Press for the Spanish Colonial Arts Society, 1983.

Wells, Eugene T. "The Growth of Independence, Missouri, 1827–1850." *Bulletin of the Missouri Historical Society* 16 (1959–60): 33–46.

Wholesale Price List. Las Vegas, N.M.: Charles Ilfeld Co., 1937.

Williams, Jerry L., and Paul E. McAllister, eds. *New Mexico in Maps*. Albuquerque: University of New Mexico Press, 1981.

Woodman, Emma. "Tin Can Trophies." *School Arts* 19, no. 4 (December 1919): 220–21.

Works Project Administration. *New Mexico, A Guide to the Colorful State*. New York: Hastings House, 1940.

Wroth, William. *The Chapel of Our Lady of Talpa*. Colorado Springs: Taylor Museum of the Colorado Springs Fine Arts Center, 1979.

———. *Christian Images in Hispanic New Mexico*. Colorado Springs: Taylor Museum of the Colorado Springs Fine Arts Center, 1982.

———. "New Hope in Hard Times." *El Palacio* 89, no. 2 (Summer 1983): 22–31.

———, ed. *Hispanic Crafts of the Southwest*. Colorado Springs: Taylor Museum of the Colorado Springs Fine Arts Center, 1977.

PHOTOGRAPHIC CREDITS

All of the photographs are by the authors, with the following exceptions:

Joseph W. Blagden: Figures 4.37, 6.13, Plate 11.

Denver Art Museum: Figure 4.15.

Roderick Hook: Front Cover, Figures 4.13, 4.16, 4.30, 5.4, 5.47, 5.84, 5.88, 5.89, 5.102, 6.12, Plates 1, 4, 5, 9, 12, 13, 16.

Index of American Design, National Gallery of Art, Washington, D.C.: Figure 4.32.

Dorothy D. and Austen Lovett: Figure 4.14.

A.R. Mitchell Museum and Gallery: Figures 4.23, 5.48, 5.78.

Museum of New Mexico, Photo Archives: Frontispiece, Figures 6.2, 6.6.

New Mexico State Archives: Figures 6.10, 6.19.

Jack Parsons: Back Cover.

Taylor Museum of the Colorado Springs Fine Arts Center: Figure 5.104.

Kate Wagle: Plates 3, 15.

INDEX

NEW MEXICAN TINWORK
1840–1940

Designed by Susan Gutnik.

The text was set in 10 point Trump Mediaeval
by Rebecca Weston with production by Don Leister
at the University of New Mexico Printing Services.
The display type was set in Zapf Chancery Medium Italic
at G&S Typesetting of Austin, Texas.

It was printed on 70-lb. Sterling Litho Matte by
Thomson-Shore, Inc. of Dexter, Michigan.
It was bound by John H. Dekker & Sons of
Grand Rapids, Michigan.